Restoring

AMERICA

The Second American Revolution

Roger Taylor

RESTORING AMERICA
THE SECOND AMERICAN REVOLUTION
By Roger H. Taylor

ISBN 0-9649605-0-8

All rights reserved. No part of this publication may be reporduced, stored in a retrieval system, or transmitted in any form or by any means without the prior permission of the publisher.

Copyright © 1995 by Technology Management, Inc.

Published by Technology Management, Inc.
1106 Rayburn Court
Mahomet, IL 61853

Printed in the Uninted States of America

DEDICATION

This book is dedicated to my mother, Marie M. Taylor, who's love, support, encouragement, countless sacrifices, and determination to have a child made this book and my life itself possible.

MY CREED

by Dean Alfange

I DO NOT CHOOSE TO BE A COMMON MAN.

It is my right to be uncommon-if I can.
I seek opportunity-not security. I do not wish to be a kept citizen, humbled and dulled by having the state look after me.

I want to take the calculated risk; to dream and to build, to fail and to succeed.

I refuse to barter incentive for a dole. I prefer the challenges of life to the guaranteed existence; the thrill of fulfillment to the stale calm of utopia.

I will not trade freedom for beneficence not my dignity for a handout. I will never cower before any master nor bend to any threat.

It is my heritage to stand erect, proud and unafraid; to think and act for myself, enjoy the benefit of my creations and to face the world boldly and say, this I have done.

ALL THIS IS WHAT IT MEANS TO BE AN AMERICAN.

CONTENTS

One	Where We Are	1
Two	Bureaucracies Gone Berserk	24
Three	It's the Stupid Economy	69
Four	Education or Indoctrination	109
Five	Crime and Circumstances	127
Six	The Sky is Falling	143
Seven	The Russians Are Coming	163
Eight	How We Got To This Point	174
Nine	The Rise and Fall of Evolution	195
Ten	Two Views of the Future	217
Eleven	The Declaration of Independence	226
Twelve	The Constitution of the United States	237
Thirteen	Restoring Constitutional Government	268
Fourteen	Survival Manual	281
Fifteen	A Few Final Thoughts	305
Appendix		310

Preface

This book is not finished. It needs more editing and some reorganization. There is a great deal more information that I wanted to put in it as well, particularly more detail as to names, dates, and places. I apologize for rushing it out in such an unpolished and incomplete form. Recent events have made it imperative that you have this information as soon as possible.

An author is not supposed to say things like the above. However, I want to be as open and honest with you as I know how. You may find some of the things in this book difficult to believe. I encourage you to check them out. There is a long list of resources in the Appendix to help you do that. You need to be sufficiently convinced of the dangers that now confront us to take action to protect yourself and your family.

I have tried to check and double check the facts set forth herein from multiple sources that I believe to be reliable. If you find any error in the facts, please send me the corrections and supporting evidence so that I may correct it.

The tragic bombing in Oklahoma City has been the best thing that has happened to Bill Clinton since he took office. His brazen use of that tragedy to promote himself and his programs are disgusting. Unfortunately, they have been remarkably successful. His approval has shot up from the low 40s well into the 50s. The anti-terrorist and no-search-warrant bills recently passed in Congress that can destroy what is left of our freedom and give him dictatorial powers were dead in the water until the bombing. Now the Republicans are scattering like frightened ants and wanting to look tough on crime and terrorism, have passed these blatantly unconstitutional laws.

We need to let our representatives know that we will not stand for this destruction of our most basic rights. Since these bills have been passed, heaven help anyone who disagrees publicly with the government. Be sure to read carefully Chapter Fourteen called the survival manual. Assuming that some government agency doesn't fabricated some excuse to throw me in jail, I plan to publish a monthly newsletter to keep you up to date on the latest power grabs and abuses by Big Brother. See the order blank in the back of the book.

Chapter One

Where We Are

"For my part, whatever anguish of spirit it may cost, I am willing to know the whole truth, to know the worst, AND TO PREPARE FOR IT." - Patrick Henry

Do you remember when:
    as a child, you could walk several blocks to the drug store for a soda in the evening and neither you nor your parents were the least bit concerned for your safety;
    farmers could plant whatever they wanted, wherever they wanted;
    the term public servant wasn't an oxymoron;
    you could go overseas and be welcomed, not attacked;
    people used the term "sound as a dollar";
    courts protected society rather than the criminals;
    you could keep almost all you earned;
    you could go through the day without unknowingly breaking dozens of laws and regulations;
    a white wedding gown meant something;
    the majority of people believed it was wrong to lie?

If you feel comfortable with the way things are today, and approve of where we are headed, you are a very rare bird. There are many valid reasons why people feel uneasy, frustrated, or concerned. I have written this book to give you information that you can use to protect yourself and your assets in the troubled times ahead. I hope that you will also act in cooperation with others to try to turn this country around and start moving in a more positive direction. There are a lot of details, facts, figures, history, and other information included for several reasons. One reason is so that you will better understand what is going on now and what will probably happen in the near future so that you can prepare for it to protect yourself and your family. Another reason for the detail is so that you can verify the information and the conclusions.

When I was in high school, the biggest problems in our public schools were: chewing gum in class, talking in class, and running in the halls. Now, the biggest problems are theft, drugs and alcohol, and assault, including rape and murder. Before leaving grade school, our children see 8,000 murders and 100,000 acts of violence on TV. The data is in. There is no question that this avalanche of violence on impressionable young minds makes kids more aggressive. They act out the violence they constantly see in solving their conflicts. Younger children can't always tell reality from fantasy. Then, there is skyrocketing teenage pregnancy and a virtual epidemic of sexually transmitted diseases, thanks largely to the liberal media, Planned Parenthood, and the sex education (encouragement) pushers. We are well into the beginnings of the New World Order that President Bush talked so much about. Are you in favor of where the people pushing and implementing these ideas are taking us? These people believe that they are the rightful rulers of the planet and are working very hard to bring us the utopia of a one-world government. Unfortunately, what they will (must) bring to pass is much more like Huxley's *Brave New World* or Orwell's *1984*. To get a clear vision of this utopia, take a good look at the Soviet Union or Eastern Europe of a few years ago, where the masses toiled for the benefit of the few elite. We are well on the way to their goal of "convergence" between communism and capitalism. Throughout history, there have always been people who were driven to set up a world wide empire. It started with Nimrod and hasn't stopped yet. Along the way, we have had such notable people as Nebuchadnezzar, Alexander the Great, Julius Caesar, Lenin, Stalin, and Hitler. The idea of a thousand-year reign (Third Reich) of a world government didn't originate with Hitler. It is as old as recorded history, and we are closer to it today than ever.

Our public educational system is a disaster. SAT scores continue to decline, and many high school graduates cannot read well enough to fill out a simple job application or add well enough to make change for a dollar. The ultra-liberal National Education Association is far more concerned about the students feeling good about themselves, even when the students cannot read or add, than they are about preparing them for a job. The NEA is also drastically rewriting history and designing curricula to program our children to accept a one-world socialist government, as in "children of the world".

Despite the recent so-called Republican Revolution, our window of opportunity to defend the Constitution and our freedom is rapidly closing. We have lost many of the freedoms "guaranteed" by the Constitution and the rest are under relentless attack. We are constantly misled by the liberal media and our own government. We are continuously being programmed to be unthinking couch potatoes, entertained into comfortable passivity by carefully concocted modern equivalents of the Roman circuses. The intention is that we will be easy to control and willing to give up our freedom in order to solve the "crises" that have actually been manufactured by, or are the inevitable result of, the actions and programs of our own government. If you think I am overstating the case, tag along with me for a while and then verify the facts in some of the many sources provided in the Appendix. Nearly all of the statements of fact herein are backed by numerous reliable sources and public documents.

We have heard much lately about the private militias being formed around the country. Most of that information has been very slanted reporting and deceptive information put forth by the government and the media. Yes, there are some extreme views held by a few people involved in these activities. Yes, there is blatant racism spouted by a few, mostly people tangentially involved, if at all. Don't misunderstand and think that I am defending any such attitudes. I think any form of racism is extremely destructive and morally indefensible. It is the cancer of history that has resulted in more human suffering and death than all other causes put together. It should not be promoted in any way. The fact is that the vast majority of the people involved in these groups are very patriotic people, many of whom have served their country in the military with honor at the risk of their lives. The question the media *should* be asking is, why are all these people so upset with the government? Why is the media not thoroughly investigating the charges against different government agencies that some of these people present? If the charges are true, we need to change some things. If they are false, then clearing the air might head off any more incredible tragedies like the bombing in Oklahoma City. Let's bring forth the charges into the public arena, as in real, thorough congressional investigations. Let's get out all of the facts and find out the truth, whatever it may be.

Congress has passed the Omnibus Anti-Terrorism Bill of 1995. Everybody is against terrorism, right? They had better be. However, this bill, like so many of these kinds of things, has little to do with terrorism and everything to do with the suspension/destruction of our "inalienable rights" supposedly guaranteed by the Constitution. This bill provides that:

1. The President or anyone under him may declare any person or group a terrorist for any reason without evidence or challenge, suspend their constitutional protections, and imprison them for up to ten years without possibility of parole.

2. It suspends Posse Comitatus, which is supposed to prevent military interference in civilian affairs, the reasons for which will be discussed later.

3. It prohibits probation or parole, even for the most minor non-violent, technical "violation". It expands the use of wiretaps, which were tightened because of blatant abuse by federal agencies in the first place. It presumes the accused is guilty until proven innocent and denies bail so the accused can be detained indefinitely, perhaps for years, with no trial. Does this sound like America or the former Soviet Union to you? Read the bill. It was bad enough in its original form and Congress was pretty reluctant to pass it. Now, thanks to Oklahoma City, Clinton has added even worse measures to it, and it was rammed through Congress because of the bombing. The lunatics that slaughtered all these people could hardly have done more damage to our dwindling freedoms.

In a confidential memo from Janet Reno to the Justice Department and to U.S. attorneys, she instituted an investigation of right-wing and fundamentalist Christian groups. The FBI will conduct investigations and surveillance of right-wing political and fundamentalist religious individuals and organizations. (Source - Alyn Denham)

Remember when Clinton said on television that Waco was an example of what might happen to people who join cults? In case you wonder who our government thinks belongs to a cult that they might use as more examples, here is Janet Reno's definition of a cultist:

> *"A cultist is one who has a strong belief in the Bible and the second coming of Christ; who frequently attends Bible studies; who has a high level of financial giving to a Christian cause; who homeschool their children; who have accumulated survival foods and have a strong belief in the second amendment; and who distrusts big government."*

How close do you came to being one of the people on their list for surveillance?

Information on these people will be sent to the Justice Department. According to her memo, in the "event of a widespread uprising, these individuals and organizations *must be* (emphasis mine) viewed as potential terrorists." In other words, the stage is set and the information is being gathered to declare political and religious conservatives as terrorists and suspend their constitutional rights. They probably are to be sent to the Civilian Inmate Labor Program camps (as in concentration camps) currently being quietly set up at abandoned military bases. A conference was held by the Justice department on August 10 - 11, 1994 in Montana with numerous agents from different law enforcement agencies. The purpose of the conference was to discuss how to link militias, pro-gun groups, conspiracy nuts, the Christian right, and anti-New-World-Order people together and to portray them all as racist, subversive, cultist, white supremacists and terrorists. These were precisely the tactics used by Hitler and Lenin with slightly different labels. (Source - *McAlvany Intelligence Advisor*, October 1994)

> *"Giving money and power to government is like giving whiskey and car keys to teenage boys."* - P. J. O'Rourke

The above comments barely scratch the surface of what is in the anti-terrorist bill and companion bills like HR-666, which throw out our constitutional protection against warrantless searches. These bills will be discussed in more detail later. There are effective ways to deal with crime and terrorism that do not require using a sledgehammer on what is left of our constitutional rights. In fact, most of these knee-jerk responses to very real threats will only make matters worse in the long run. If these blatant attempts to throw out the Bill of Rights don't terrify you, then you probably don't know the history of Nazi Germany, the Soviet Union, or Communist China very well. The parallels between what happened in Germany in the 1930s and what is happening in America now is amazing and frightening. Remember, Hitler was democratically elected and then slowly eliminated freedoms, especially gun ownership, because of various "crises" until he was ruling by decree. He tried to make everything look legal and necessary so there would be no widespread objections.

The same kind of "decrees" are now on the books in the U.S. as presidential executive orders, and await only the President to declare an emergency, real or otherwise. When (not if) he does, our Constitution and the Bill of Rights will be gone forever. Both Congress and the Supreme Court will be helpless to do anything about the situation under the provisions of these orders. They should be declared unconstitutional right now, while we still have that opportunity, but apparently nobody with the right "standing" has challenged them.

Many of our regulating agencies are running amuck. Several years ago, Congress declared that the Bureau of Alcohol, Tobacco, and Firearms (BATF) was a renegade agency that was out of control, but did nothing about it. That was before Waco and numerous other tragic "incidents". The BATF is not unique in this respect. Randy Weaver's wife was shot in the head and killed by an FBI sniper while standing on her porch holding her baby. Weaver's fifteen-year-old-son was shot in the back and killed by marshals as he ran toward his father. The FBI originally claimed that the murder of Weaver's wife was an accident. They said that their highly trained sniper (who could consistently hit a quarter at that range) was actually shooting at her husband. The story has changed several times. We now know that FBI supervisors illegally changed the standing orders of engagement to an order to shoot to kill anyone on the property – which the FBI did. These deaths were cold-blooded murder. Finally, the facts about the cover-up, lying, and misconduct are starting to be made public. Unfortunately, the same kind of cover-ups occur almost constantly in many government agencies. We probably never will know the whole truth about Waco, and it is unlikely that any government agent will be charged with murder in either situation.

With the media's help, many government agents think they can hoodwink the public and cover up almost anything. Does this sound possible in America? It is becoming an all too common occurrence. There are more such incidents described in Chapter Two. The same trigger-happy agents who ordered the Weavers shot, rejected the FBI's own psychology team's plan to get David Koresh to surrender peacefully, and did just the opposite of what they told them needed to be done to end the standoff quickly and easily.

The recent seizure laws and court decisions have opened a Pandora's box that has cost thousands of innocent citizens everything they have, including their lives. We have, in effect, told law enforcement agencies that they can go out and confiscate anything they want to, without any proof or even evidence of wrongdoing, and *profit from what they confiscate!* Regardless of the innocence of the parties involved, it is very difficult to ever get the property back. Anyone who would vote for these kinds of laws that actually provide financial incentives for the police *and the courts* to raid innocent people is either extremely ignorant of human nature and history, or has a very sinister hidden agenda.

The Republicans have caved in to requests from the "Justice" Department to drop a provision in the Contract with America that would require the government to notify citizens that they are under investigation. Gingrich and Dole also agreed to avoid the repeal of the assault weapons ban.

I am a strong supporter of law enforcement, particularly local law enforcement. The thin blue line, as it is called, is all that stands between us and anarchy. But what do you do when the guys with the badges are focused on a mission and don't care who they trample in the process? What if they know they will be protected by their political superiors (as in Waco for example), no matter what they do? The Swat team/Gestapo/KGB mentality, which is permeating more and more of our federal agencies, is *far* more dangerous to innocent people than are the crooks these agencies are supposed to be protecting us from.

8 - Restoring America

A case in point. Donald Scott was in bed and heard his door being broken down. He picked up a revolver and went to investigate. As he entered his living room, the door was battered down and a sheriff's deputy ordered him to drop his gun, which was at his side and pointed at the floor. He did. The sheriff's deputy then shot him twice in the chest, killing him. His crime? None. They were there because he was an heir to a European fortune and owned a $5 million ranch. They raided the ranch to see if they could find anything that would let them seize the ranch for forfeiture for their agencies. How do we know that? Evidence turned up that the raiding party had gotten appraisals of his property and the sale prices of nearby property before targeting his ranch. They found nothing. No one involved in the raid was charged with the murder of Mr. Scott nor any of the other crimes and civil rights violations they committed.

By the way, did you know that if someone knocks on your door and you open it to find a police officer there that you have, under the current arrangements, unknowingly thrown away any rights you have to require a search warrant? He can now enter your house for any reason he wants to, without your permission.

Bye-bye, Fourth Amendment. Just by opening the door to see who it is, you are, supposedly, giving implied consent for these people to come in and do whatever they want to. Of course, if you somehow knew who it was and what they wanted, you could refuse to open the door. In that case, they would just break it down anyway.

Then there is the FDA, the IRS, the FTC, the EPA, OSHA, and even the Post Office – ad nauseam. Even agencies who have no conceivable enforcement needs are getting their own stormtroopers. Why? American soldiers are being asked whether they would fire on U.S. citizens if those citizens refused to surrender their guns. Why? Please read the following statement very carefully and thoughtfully. Remember it as you read the rest of this book.

> *"In Germany, the Nazis came for the Communists and I didn't speak up because I was not a Communist. Then they came for the Jews, and I didn't speak up because I was not a Jew. Then they came for the Trade Unionists, and I did not speak up because I wasn't a Trade Unionist. Then they came for the Catholics, and I was a Protestant, so I didn't speak up. Then they came for me...By that*

time, there was no one left to speak up for anyone." - Rev. Martin Niemoller (1945)

The march toward an American police state continues. The House of Representatives voted 289 to 142 to allow law enforcement officials to break into homes and search them without a warrant. This Bill (HR-666) was pushed through by the Republican leadership (220 Republicans and 69 Democrats voted for passage). It is a total violation of the Fourth Amendment to the U.S. Constitution, which outlaws unreasonable search and seizures. Once again the Republicans show their true colors!

As Benjamin Franklin said after signing the Declaration of Independence, "We must all hang together, or we shall all hang separately." Both liberals and conservatives who understand what is being taken from us, piece by piece, must join together to put a stop to it. The door is rapidly closing.

There are bills currently pending in both houses of the Montana legislature to secede from the United States. These people are not far-right, white supremacist, racist lunatics. They are very rational people who are concerned about the growth of federal power, the abuse of that power, and the growing intrusion into personal freedom. They have very valid reasons for their concern.

HR-737, introduced by Rep. Reynolds (D-IL) would make the manufacturer, importer, or dealer of a handgun or an "assault weapon" liable for damages that result from the use of the firearm. Nobody could afford to stay in the business. Obviously, that is their goal. Is this any different from making automobile manufacturers or dealers liable for damages whenever someone has an accident? It is nonsense, but dangerous nonsense, as it would drive all American manufacturers out of business and stop all imports. This would eliminate all but private sales of existing guns, resulting in a huge black market that could end up nearly as lucrative as drugs.

THE ECONOMY

"My friends, we are only a few years, at most, away from an historic crash which will cripple the American economy and erode confidence in the legitimacy of our political system." - Lawrence B. Lindsey, Federal Reserve Board Governor

10 - Restoring America

Then there is the economy, or what's left of it. We are sitting on the biggest speculative bubble in history, thanks to the Federal Reserve System, which is owned by large private banks, creating billions of dollars out of thin air to try to bail those banks out of their bad loans. We are walking along a precipice and will soon go over. We are at a time in history where many cycles of history and natural phenomena are all coming together in a way that is unprecedented. There will be a number of years of great instability with the breakdown of governments (already in progress) and economic chaos. The Dow average will go to 1,000 or below, wiping out most retirement systems and personal investments. The federal government, which caused the problems in the first place, will do exactly the wrong things, just as it did during the Depression. It will enforce Draconian measures and consolidate power in Washington with massive central planning and control. This will only make matters much worse, as the total failure of the central planning and control of socialism/communism clearly demonstrates. Unfortunately, that has been the natural response to crisis by those in power throughout history. It was World War II that brought us out of the Depression, not Roosevelt's socialistic policies. Actually, by the time many of the policies were put in place, the economy had begun to recover, and it fell apart again because of them. The disastrous long range effects of Roosevelt's programs have yet to be felt in their full power.

In the 1970s, one dollar would buy 300 yen worth of TVs, cameras, or Toyotas. Today, it will buy 80 yen worth. When Clinton took office, one dollar would still buy 120 yen. That is a decline of one-third in just two years. A few years ago, we were the largest creditor nation in history. Virtually everybody owed us money. We were the world's banker. Today, we are the largest debtor nation in all of history. We are technically bankrupt with no legitimate way to pay our bills, and what productive base we have to pay even the interest is fast eroding. It is no wonder foreign and even American dollar holders have been dumping these questionable pieces of paper.

Eighty percent of our Gross Domestic Product is now in service industries. That means eighty percent of our work force is shining our shoes, fixing our cars, filling our teeth, and mostly shuffling government required paper. All these people are producing nothing. Only twenty percent of the work force is actually producing anything that can be sold elsewhere to bring back U.S. dollars, and nearly all of that is consumed here. That is why our horrendous trade deficit cannot be tamed. No economy can survive long with an 80/20 ratio. For the past twenty-five years, real wages, discounted for inflation, have steadily dropped as our production jobs have gone overseas. When all we do is service each other, a good chunk of what we earn gets confiscated and poured into the black hole of government programs that produce nothing. Another large chunk of it goes to other countries either as foreign aid or to buy our high-tech toys that we can't produce at a sufficiently low cost ourselves. Virtually nothing goes into savings to fund new businesses and modernize older ones. Thus, the downward spiral continues. Can we do anything about it? Yes, but as usual, the politicians are doing mostly the wrong things.

NAFTA and GATT were designed by the Trilateral Commission as two giant steps towards creating the three giant trade blocks that is its stated purpose in life. These three major trading areas are designed to lay the groundwork for the new world socialist order by generating all the entangling regulations inside the groups. Once they are functioning, it is their plan to easily merge them into one world trading block. These treaties are a major Trojan horse which will not only override all federal and state laws, but the Constitution as well. Our courts have no choice but to enforce the decisions of these secret foreign bureaucracies.

Foreign businessmen are delirious with joy over what they can do to us with these treaties. Both the Japanese and the Europeans have a whole list of "grievances" against American businesses ready to be filed that these businesses must defend themselves against or be fined out of existence. However, the cost of defending themselves will cause them to go broke anyway, which is exactly the reason for filing the grievances to start with. The European Economic Community has a whole list of proposed sanctions against the U.S. which have no purpose other than to keep us from competing with them. We have been suckered again, as usual. Brave New World, here we come. One of the things that Ross Perot was right about was the horrendous cost of NAFTA. The multinational corporations wanted these treaties so they could export our jobs to low wage countries with few or no regulations and import the products back with no duties. Bye-bye, production jobs. The international banks and major brokerage firms wanted NAFTA to bail out all of their bad loans and Mexican stocks and bonds they bought themselves and sold to their customers. Both they and our elected officials in Washington knew the peso was being artificially supported by massive loans and creative bookkeeping until the Mexican presidential election. It happens with every election down there. That is why there was such a big rush to get NAFTA and GATT through before the new Congress was sworn in. Not only was there the possibility that some of the new Republicans might not go along with this disaster, but more importantly, the peso could crash at any moment and scuttle the treaty.

Who made it possible to get these treaties approved? Bob Dole. Which two individuals strongly supported both the treaties and the bailout when the peso did crash as expected? Newt Gingrich and Bob Dole. Where does much of their PAC and other campaign money come from? Multinational corporations, banks, and brokerage houses. Any questions about who they really represent? It certainly isn't working Americans. Not only will the bailout cost us the twenty billion (our tax money) dollars pledged to support the peso, but the treaties and the bailout are seen by other countries as a clear sign that the dollar is tied to the southern peso of a totally corrupt Mexican government and the northern peso of Canada. Canada is even deeper in debt per capita than we are because of the same kind of socialist programs, such as health care that our government is trying to foist on us. Tying our dollar solidly to these two sinking currencies through NAFTA was the trigger for the big run on the dollar. We are a soggy life preserver, now firmly tied to two sinking ships. The recent run up in the dollar is due to artificial manipulation by the central banks to try to help the Japanese economy. It can't last as our continuing trade deficits ship billions of dollars overseas every month.

Initially, there was a net gain from NAFTA as American corporations shipped production machinery south to open new factories (there go more jobs) and Wal-Mart and others stocked up their stores. These are the numbers that the supporters of NAFTA still quote. The gain quickly became a net loss and the stores were virtually empty before the peso crashed. Now it is much worse.

The trade imbalance and net job loss is becoming very embarrassing to the supporters. Do you really want to re-elect the people who are tying you to foreign (socialistic) bureaucracies, failing economies, and bringing you the second Great Depression?

> *"Everybody, soon or late, sits down to a banquet of consequences."* - Robert Louis Stevenson

WELFARE

Hardly anyone would argue that our welfare system is not a sorry mess. It is using and abusing the recipients almost beyond description mostly for political purposes. It is draining the economy and creating horrendous social problems for everyone. Federal welfare is unconstitutional to start with, clearly counterproductive, and has to be scrapped, not fine tuned. The details are in Chapter Four.

> *"Democracies cannot exist as a permanent form of government; they will only exist until the people find out that they can vote money for themselves from the treasury and until the politicians find that they can distribute that money in order to buy votes and perpetuate themselves in power. Hence, democracies always collapse over weak fiscal policy to be followed by a dictatorship."* - Alexander Tyler, British historian.

> *"History makes it plain that unless restrained, government proliferates to a point where its cost bankrupts the people at the same time it robs them of their freedom."* - Ronald Reagan

We are on the brink of that economic collapse.

Americans in particular tend to be optimistic and believe that somehow everything will work out all right. That is a good and necessary attitude when building a nation. However, it is deadly when confronting very intelligent, well financed, highly-organized people committed to the destruction of U.S. sovereignty and to our submission to a one-world socialist government. An ostrich with his head in the sand may feel safer because he cannot see the approaching enemy, but that only sets him up to have his rear end kicked between his ears.

> *"But the great majority of mankind is satisfied with appearances, as though they were realities. ... And they're often more influenced by things that seem, than by things that are."* - Machiavelli

This is why the far-left-liberal media has been so successful in misleading people and also why over forty percent of the people would vote for "slick Willy" again. It is necessary for the government to hide the truth these days, with the help of the liberal media, or they would have a revolution on their hands now. There will be a revolution! What we need is a true revolution of ideals that take us from what is prevalent today back to the foundation principles that made this nation great. The little Republican "revolution" that we have going on now is just a bear-market-rally "blip" in an ongoing downward plunge. The excuses we hear from the politicians about why we can't do this or that, or what they want to do to "solve" the crisis that they have created are not only absurd, but dangerous to our freedom and economic well-being.

The big government, central planning, and socialistic principles that we have been following for the past sixty years just don't work. They are fatally flawed in their philosophical foundations, because they ignore human nature. For over one hundred and fifty years, this nation grew to become the greatest economic power the world has ever known by mostly following the principles on which it was founded. For the last sixty or so years, we have slowly turned away from them, and we now face the worse economic, moral, and legal crisis in our history. Socialism and socialistic programs don't work. The communists have slaughtered over one hundred million people trying to impose socialism and make it work. Look at the condition of their countries today. They have had the greatest social experiment in history, and it has resulted in unparalleled slaughter and suffering. You don't need to be a student of history or human motivation to realize that demotivating socialistic concepts and centralized power are a disaster. Our founding fathers knew that. Read the Federalist papers and the Anti-Federalist papers.

If you don't think that we have swallowed the socialist agenda wholesale, here is what the man who wrote the Socialist Party platform in 1932 had to say.

> *"The American people will never knowingly accept Socialism, but under the name of Liberalism, they will adopt every fragment of the Socialist program, until one day America will be a Socialist nation without knowing how it happened... The United States is making greater strides toward adopting Socialism under Eisenhower than under President Franklin D. Roosevelt."* - Norman Thomas, *Two Worlds*

Before his death in 1970, Thomas stated that the U.S. had adopted every plank in his socialist platform. He ought to know. (Source - *McAlvany Intelligence Digest*, October 1994)

Our liberal-evolution-based educational establishment has stopped teaching real history and these kinds of principles and substituted instead politically correct indoctrination. We have to acknowledge the emptiness and devastation of high-sounding theories and go back to what actually works. Any attempt to just tweak the faulty system is doomed to fail. It is based on false and failed philosophies.

COUNCIL ON FOREIGN RELATIONS AND TRILATERAL COMMISSION (CFR/TLC) PLANS FOR YOUR FUTURE

As mentioned above, the window of opportunity is open for only a short period, for both sides. The "other side" knows it all too well. David Rockefeller, who founded the Trilateral Commission, speaking to the Annual United Nations Ambassadors Dinner on September 14, 1994, stated: "But this present 'window of opportunity' during which a truly peaceful and independent World Order* might be built, will not be open for long. Already there are powerful forces** at work that threaten to destroy all of our hopes and efforts to erect an enduring*** structure of global cooperation." ****

16 - Restoring America

For those of you who might not be familiar with the buzz words these people use, *World Order means new one-world socialist government. **These powerful forces he is talking about are patriotic Americans who want to preserve the present U.S. Constitution and U.S. sovereignty. *** Enduring means no one will have any weapons or hope of overthrowing it as in the days of the Roman Empire. **** Global cooperation means bureaucratic dictatorship which will ultimately be a one-man dictatorship, as history amply demonstrates. Can these people slam dunk a one-world-socialist government on everybody by the year 2000 as they have been planning for a very long time? Only if two things happen. One, nearly all the guns in the U.S. must be registered and then confiscated, which is beginning to happen already. Two, the Constitution must be replaced or fatally weakened. Try this "news story" on for size.

DATELINE: Philadelphia, Pennsylvania. 1997. The Conference of States adjourned today after passing a new Constitution for the U.S. This conference was called, ostentatiously, for the purpose of changing the Tenth Amendment to strengthen the powers of the states and to pass a balanced budget amendment. However, like the conference of states in 1787 that was only authorized to tweak the Articles of Confederation, they threw out the whole thing and started over. This new Constitution is virtually identical to the Newstates Constitution drawn up by a consortium of the Ford Foundation, the Rockefeller Foundation, the Carnegie Foundation, and the Rockefeller-Spellman Foundation between 1965 and 1974. This proposed constitution was prepared for the constitutional convention that the Rockefellers tried to convene in 1975. Our new constitution, which will take effect on January 1, 1998, has been created to provide a more streamlined form of government better suited to the rapidly changing times we now live in. It provides for: 1. The dissolution of the present state governments and boundaries as we know them and their replacement by ten new states of approximately equal populations, except Hawaii and Alaska, which will remain as they are. These streamlined states will be able to carry out new federal programs much more rapidly. 2. The powers of government will be concentrated into the hands of eleven individuals, appointed by the President (who will be elected for nine years) in order to provide for greater efficiency. These are the Overseer, the National

Where We Are - 17

Regulator, the Principle Justice, the Watchkeeper, the Intendent, the Public Custodian, the Chairman of the National Planning Board, and the Chancellors of External Affairs, Internal Affairs, Legal Affairs, and Military Affairs. 3. It guarantees freedom of expression, communications, movement, assembly, or petition unless the President declares an emergency (Article 1, Section 1). 4. It establishes federal gun control and licensing for individuals who meet the necessary qualifications and needs (Article 1B, Section 8). Other provisions provide that all political parties will need the approval of the Overseer. All spending on elections will be controlled by the Overseer. States which fail to carry out the mandates and regulations of the federal government will lose part or all of their funding which shall be supplied entirely by the federal government. These and other provisions will be explained in detail when copies of the new Constitution are printed and distributed to the media.

Absurd? Fantasy? Impossible? Wrong! Most of the number of states necessary to the convening the "Conference of States" have already passed a resolution to this effect in one or both houses. The conference was set for October 20 - 25, 1995. It has been postponed, fortunately, due to the current political situation and a lot of negative publicity. Bob Dole and Newt Gingrich were going to be on hand to endorse the Constitutional Convention, and also to make it constitutionally legal when the Conference of States is officially converted from the COS to a Constitutional Convention that the CFR and their new "front" groups have been striving for. Newt has stated that he is "working with the leaders of the Conference" to this end.

In most of these resolutions there is language that supposedly limits what the conference can do, but such language is irrelevant, as the planners of this disaster are well aware. The structure of this conference is deliberately being set up just like the first Con-Con in 1787. William Burger, former Chief justice has said: "There is no effective way to limit or muzzle the actions of a Constitutional Convention. The convention could make its own rules and set its own agenda." Most constitutional scholars agree that the conference could vote to become a constitutional convention just as they did in 1787 and nobody could do anything about it. Would they? This is exactly what they are planning on doing. Delegates to this conference will not be elected, but will be hand picked by the Establishment.

I wonder which of these positions (overseer, regulator, or whatever) corresponds to Hillary's more correct occult title of Gatekeeper in her communistic health care allocation and population control plan. Actually, no health care plan in any Communist country ever went as far in terms of control and allocation as hers did.

Sun Tzu (500 BC), the favorite warfare strategist of the communists, said: "All warfare is based on deception." There is an undeclared war on the U.S. Constitution. Actually, these people have made their intentions quite clear in their own writings and statements, so maybe it isn't as undeclared as it might seem. It is just that the American public doesn't know what is going on, thanks to our far-left-liberal media. Everything about this whole affair is classic Orwellian doublespeak. The conference is supposedly being called to enhance the power of the states. Sure it is. Anyone the least bit familiar with the people and groups that are organizing and pushing this scheme knows better. The carrot that is being dangled in front of state legislatures to convene this nightmare is more money from Washington, which, in most cases, is actually unconstitutional in the first place. Instead, this conference has been very cleverly designed to *eliminate* the states, along with our most cherished freedoms.

If you want more details on this attempt to eliminate your freedom once and for all, send $2.00 to the *McAlvany Intelligence Advisor*, P.O. Box 84904, Phoenix, AZ for a copy of the April 1995 issue, which has most of the gruesome details and references. You may also obtain a copy of the Newstates Constitution, which you may soon be living under, from Liberty Library, 300 Independence Ave. S.E. Washington, D.C. 20003 for $3.00. If this all sounds too extreme to be true, I don't blame you. Not in America, surely. So, don't take my word for any of it. Certainly, don't take the disinformation (the current euphemism for lying through one's teeth) of the media and the government for anything resembling truth. Deception is rampant in all sections of society, from the highest level of government all the way down. The people who control and finance the foundations mentioned above have your future planned for you. To find out what they are really after, get a copy of the Newstates Constitution *that they wrote!* Read it carefully. Notice the words missing or phrases added that sound like our Constitution. Words have meaning, especially in legal documents. Our founding fathers had no illusions about the probability of men abusing the power of government for their own ends. They went to great lengths to limit and sharply define the limits of that power.

For instance, this new constitution states that: "No property shall be taken (by the government) without compensation." What's wrong with that? Plenty. They deliberately left out the vitally important word "just" on purpose. If the government wants to take someone's property, who determines the price (if any) under this new constitution? The government! Remember the provision mentioned above about the "rights" of free speech, etc.? It clearly shows the differences in philosophy of the people who want to give us their new constitution and the founding fathers.

The people who crafted the Constitution firmly believed that men had certain "inalienable rights" to start with. Nobody "allowed" them to have them. They put together a foundation document that grudgingly gave a central federal government the power to exercise a very limited, carefully structured authority to provide for more effective defense and efficient commerce. They went to considerable effort to reserve all those inalienable rights for the people and protect them from government encroachment.

The people who wrote the Newstates Constitution have just the opposite view. To them, it is the government that has all of the power and authority, and grants to the citizens (serfs) certain privileges as long as they behave themselves and do as they are told. Remember the provision mentioned above concerning free speech? Notice that you will have free speech unless the President decides that there is an emergency, such as a rebellion against the government's tyrannical rule. You have freedom of speech, or whatever, not as a right, but as a privilege, subject to the whim of the President. Again, don't take my word for it. Get a copy of it and compare it carefully with our current Constitution.

Just in case the people who want to radically change or eliminate our Constitution don't manage to get an entirely new constitution at COS, they have plans B, C, D, and E. They always do which is why our freedom is slowly dying. Plan B is as follows: If enough of the hand-picked delegates somehow refuse to go along with a new constitution, they will then try to weaken Article V (amending) dramatically so that they can easily piecemeal the Constitution into what they want. They will also drastically revise the Tenth Amendment to strengthen the federal government's power, even though the conference was called to do just the opposite. Plan C is to pursue a constitutional convention under the resolutions already passed by twenty-nine states calling for one. Jesse Helms, of all people has introduced a resolution for just such a convention. He and the cosponsors think they can limit such a convention to just passing a balanced budget amendment. They cannot limit it no matter what the resolutions say. Please Jesse, don't do this to us. It is just too dangerous. The whole drive so far behind any kind of a constitutional convention has come from the Rockefeller crowd. They want to scrap the Constitution and pass the Newstates Constitution that they wrote.

Even if you have never contacted any of your state legislators for any reason, do so. Drown them in calls, faxes, mail, e-mail, anything you can think of. Ask them if either the COS or the CON-CON resolutions have been passed in their chambers. If not, ask them to do everything in their power to make sure they are not. If they have been passed, they must be rescinded before it is too late.

CLINTON'S AGENDA

Four major goals are on Bill Clinton's agenda:
1. Impose socialized medicine as a major step towards complete government control of the people;
2. Require some sort of identification smart card to facilitate this control;
3. Pass NAFTA and GATT to begin subordinating our national sovereignty to foreign bureaucrats;
4. Disarm the American people.

So far, numbers one and two have been defeated, but both the post office and the IRS have come up with their own versions of such a smart card as backups. Number three has been successful thanks to Bob Dole. Significant strides have been made towards number four and the Oklahoma bombing is being used to further that goal in a big way, even though guns had nothing to do with the bombing.

CYCLES

If our government leaders had any knowledge of real history (or cared), they would know that we are in the beginning years of an unprecedented confluence of a number of major physical and historical cycles. These include some as long as five hundred years. They should be preparing for the results of these cycles, but are far more interested in their own welfare, power, and influence than in the country's welfare. These cycles indicate substantially greater numbers and magnitudes of earthquakes, volcanic eruptions, breakdown of governments (especially central governments) into regional and tribal groups, lawlessness, breakdown of economies and money systems, drought, wars, confusion, and chaos. It has already started and will get worse during the next decade. We will see massive riots in our inner cities.

During 1989 - 1991, the sun's motion relative to the center of mass of the solar system was retrograde. This happened before in the 1630s and 1810s. Each time, the result was a dramatic increase in volcanic eruptions and extremes of climates. The change in gravitational forces throughout the solar system stresses the continental plates and causes more frequent and more violent earthquakes. The chances of serious droughts are increased.

We are at the end of a five-hundred-year civilization cycle that normally sees the breakup of existing nations and civilizations into smaller units with larger numbers of wars and conflicts. Have you looked around the world lately? We are at the end of a sixty-year business cycle (Kondratieff supercycle) that proceeds from depression to savings, growth, prosperity, credit expansion, credit abuse, collapse and depression. We are at the bottom of the eleven-year sunspot cycle, which has a profound effect on activities on the earth. Over the next fifteen years, we will see virtually everything we take for granted changed or disrupted. Asia will become the dominant economic area. Equatorial warming, coupled with polar cooling, will dramatically affect the weather. El Nino will affect the jet stream causing it to be vertical (North-South) much more often than horizontal. This is already causing unusual flooding in some regions and drought in others. Is our government preparing for this? Of course not. They are too interested in gaining control over every aspect of our lives.

"There is a high probability that the U.S. will be involved in a major war during the next ten years. President Clinton has many reasons to desire a war, whether consciously or otherwise. His repeated ultimatums to leaders of other countries (seven by last count; ...) and his willingness to dispatch troops to numerous locales support this conclusion. His Russian counterpart, who enjoys an even lower popularity, has also been quick to adopt military solutions. In a dramatic yet overlooked indication of Clinton's support for armed solutions, Paul Goble, a Russian expert at the Carnegie Endowment for International Peace, stated that 'Yeltsin's only supporters (in his military invasion of Chechnya) appear to be ultra-nationalists like Vladimir Zhirinovsky and the Clinton Administration." - Robert Prechter, *The Eiliott Wave Theorist,* January 6, 1995

"Joshua S. Goldstein, who has made an exhaustive study of war cycles, pinpoints the year 2000 for the beginning of the 'danger zone' for great power war." - Julian M. Snyder, *The Armageddon Letter,* August 1994

"..Clinton will not sit on his hands. He knows the one thing that is sure to make people rally around a president is war..."Couple this with the fact that in Iran the economy is a mess and Iranian rulers need a war, and in Iraq Saddam Hussein needs a war...." - Richard Maybury, *Early Warning Report,* January 1995

"China's political situation is getting shakier by the day. Paramount leader Deng Xiaoping is ninety and ill, and no one knows who will replace him when he dies. China has a long and recent history of feudalism, so my guess is the country will revert to that, breaking up into dozens of small states at war with each other. The same as what's coming in Russia " - Richard Maybury, *Early Warning Report,* January 1995

Throughout this book, and particularly in the "survival manual" in the back, I will tell you how you can try to insulate yourself, individually, as much as possible from the coming economic earthquake. However, there is little we can do to stop the destruction of what is left of our constitutional rights, unless we act collectively and massively in the voting booths and political organizations, *now*. Thus, this little treatise. It is not intended to be exhaustive; that would take a set of encyclopedias. I hope that it *will* sound an alarm, point you to the kind of references and continuing information that you need to protect yourself, and motivate you to do something while we still have the opportunity. We are going to have a revolution; it has already started. The question is, will it be a revolution of ballot boxes, or of bullets and bombs? The choice is up to us. If we don't have the former, we will have the latter, and soon.

There are no footnotes in the text for several reasons. For one thing, they are very distracting. For another, some of this information has been collected over thirty plus years. I don't remember all the sources precisely, particularly those I found prowling among the dusty stacks of the University of Illinois library over the years. However, nearly everything I state can be verified somewhere in the material listed in the Appendix, usually from several reliable and proven sources.

The next six chapters include a more detailed continuation of the themes mentioned above, as well as some suggestion on how you can insulate yourself and your family from the results of our government's folly as much as possible. Don't get discouraged about all the problems. We can change things for the better if we want to do so and are willing to work together. Those who prepare for the worse will come out the best.

Chapter Two

Bureaucracies Gone Berserk

"The more corrupt the state, the more laws." - Tacticus, Roman General

"The American public does not have the knowledge to make wise health care decisions...FDA is the arbiter of the truth....Trust us. We will tell you what's good for you." - David Kessler, MD. Commissioner for the FDA on the Larry King Live Show.

Sure, Mr. Kessler. Your connections to the large drug companies and the "conventional" medical establishment, along with your demonstrated determination to wipe out alternative health care will not cloud your omniscient objective decisions one bit, will they? His statement pretty well sums up the *"I am god; I will take care of you"* attitude of the liberals in general and high-level bureaucrats in particular. No agency in government is permeated with that attitude more than the FDA, but it is present in all regulatory agencies to a frightening degree. The idea of our nanny government is that we are all a bunch of stupid sheep. They must feed and care for us, make our decisions, defend us from ourselves and fleece us in the process as their rightful due.

I was about one-third of the way through the process of writing this book when the Oklahoma bombing occurred. Because of it, I rearranged the order of the chapters and moved this one forward several chapters. I have been aggravated, but not surprised, by the media's slanted reporting, pejorative labels, and deliberate smearing of a number of people and groups, especially the private militias. Yes, there are some kooks in these groups, as in most organizations. Any group that is opposing certain actions by any government is going to attract some frustrated and/or angry people. However, most of these people are honest, patriotic, concerned citizens who are active in their communities, strongly support the constitution, and believe in law and order. They have some very valid concerns that lead them to spend their valuable time and money preparing to defend themselves and their (and our) constitutional rights against their own government, if that should become necessary.

Why would intelligent, knowledgeable people believe that such extreme measures might become necessary? Well, they have history on their side, for one thing. They also hold the same basic concepts of freedom and mistrust of centralized government that our founding fathers held. They have a lot of recent events and examples to bolster their suspicions.

Nothing can possibly justify or excuse the senseless tragedy in Oklahoma. We can only hope that somehow, something good can eventually come from it. A good start would be an honest, open, thorough investigation by Congress, which is responsible for the oversight of these agencies, into the many charges that have been made against them. Waco, which is only one of many incidents, is a festering sore on the nation's soul. We need to deal with them fully, including special prosecutors since the Justice Department is responsible for many of these crimes.

BUREAU OF ALCOHOL, TOBACCO, AND FIREARMS

Some people refer to the BATF as the Bureau of Atrocities, Terrorists and Fanatics. With apologies to the many honest, dedicated people in the BATF, that is an obvious overstatement. However, it is the opinion of growing numbers of people that now number in the millions. Do they have a valid reason for that opinion? About fifteen years ago, a friend of mine was a BATF agent. One day he mentioned that he had resigned. He was reluctant to talk about it, but I found out that he had no other job lined up and no idea what he was going to do to support his family. When I asked why he had resigned, he told me that the agency was becoming more and more illegal and demanding in the harassment of gun dealers and pawn shops. He could not in good conscience do what they were demanding anymore. That was fifteen years ago and things have become much worse since then. Let's take a look at some recent events.

MONIIQUE MONTGOMERY. Early one morning in July 1994, the BATF broke into the home of Monique Montgomery. Supposedly, they were looking for drugs. They found none.

They burst into the bedroom where she was sleeping with their guns drawn. Waking up, startled, with her bedroom full of masked men, she did the most natural thing in the world. She reached for a gun she had lawfully purchased to protect herself from home invasion. She never made it. They shot her four times as she lay in her bed. She survived, somehow, but after a long time in the hospital, she is still confined to a wheelchair. No charges were filed by anyone, because she had done nothing wrong. She is one of dozens of people who have been gunned down, most of them killed, by the rampaging federal law enforcement agencies who forcibly break into people's homes in clear violation of the Constitution. This must be stopped. Yet the Republican Congress has voted to extend the no-knock powers of these agencies, as Clinton had requested. If the Supreme Court doesn't have sense enough to overturn these laws, our only hope is to replace the current politicians. You can't outgun the government. Don't feel it can't happen to you because you obey the law. Some of these people have been terrorized or even killed when the stormtroopers got the wrong house by mistake. If these were one or two isolated incidents, it might not be so scary, but it has become standard procedure for most agencies, especially the BATF.

LOUIS KATONA, III. Mr. Katona is a police officer who liked to collect machine guns, perfectly legally. The BATF doesn't like such people. In May 1992, they raided his house. During the raid, they threw his pregnant wife against the wall so violently, without any reason, that she miscarried what would have been their second child. They seized over $100,000 worth of perfectly legal guns and did everything they could to keep them. Katona sued the BATF to get his property back. In response, the BATF tried to indict him to intimidate him into dropping the suit so they could keep his property. He refused. The judge threw out the phony charges of the BATF. Two city policemen who refused to lie about Katona in court were fired for not "cooperating" with the BATF persecution. Eventually, after a long delay and a lot of legal expenses, Katona got his guns back.

JOHN LAWMASTER. Mr. Lawmaster has never been accused of any crime. He has done nothing to attract the attention of law enforcement agencies. One day, he returned to his home to find both doors smashed in, his gun safe broken open, drawers and cabinets emptied all over the floor, ammunition thrown around and his home completely trashed. Malicious vandals? No, the BATF had paid him a little visit. According to witnesses, several vehicles pulled up in front of his home and about thirty BATF agents piled out with weapons drawn. They surrounded the house. They did not announce themselves or make any attempt to serve a warrant. They smashed in the front and back doors. They broke open his gun safe and dumped ammunition on the floor. They examined the guns, particularly an AR-15, the only thing mentioned in their search warrant. Finding it perfectly legal, and having no excuse for their raid, they then proceeded to remove clothes from hangers and throw them on the floor. Dresser drawers were dumped on the floor. A bed was overturned. A lock box was battered open. A desk was emptied on the floor and files left scattered around the room. Ceiling tiles were poked out. Although they had found the object of their search, and verified that there was no violation, they proceeded to break into a shed and turn it into a shambles as well.

Next, they started to break into a motor home near the house. They stopped only when a next-door neighbor identified it as his. Despite the fact that they had no warrant or legal basis for doing so, they forced the neighbor to open his motor home. They grilled the neighbor about whether his neighbor's AR-15 had an automatic sear. He told them he didn't know. Noticing that the neighbor happened to be wearing an NRA hat, the agent said: "You're one of them", implying that he was some kind of enemy. He was then threatened with seizure of *his* property if he didn't cooperate with the search. Finally, the agents left. These people, who are supposed to be so concerned with preventing crime and keeping guns out of the hands of criminals, walked away leaving the doors open, with guns, ammunition and his other property readily available to kids or anyone else who wanted to walk off with it.

Since when can someone's property be seized if the police don't like the way they are cooperating with the search of someone else's property? Well, similar things are happening all over the country. When Mr. Lawmaster attempted to find out what triggered the raid, the BATF had the affidavit that was submitted to get the search warrant issued sealed. For a long time he had no idea why his house was trashed, and the authorities refused to give him any information. When he tried to get the BATF to pay for the damages, they laughed at him. Was this raid a mistake? No, they came to check out the AR-15. It turns out that a woman he had just divorced went to the BATF and lied to them and told them that Lawmaster had illegally converted the AR-15 to a fully automatic weapon, in hopes of causing him trouble. She did.

The BATF made no attempt to verify her lie before wrecking Lawmaster's home. Why didn't someone just come to his home when he was home and ask to see the gun? Ask the BATF. Did they have ANY evidence of a crime or reasonable cause for the warrant? No! Did they even attempt to serve the warrant? No! Nowadays, if these agencies even bother with the legal "niceties" of obtaining a warrant, they frequently don't try to serve it lawfully. They just break in and all too often, someone dies unnecessarily. The warrant, if they have one, is only an excuse, a legal backup in case someone questions what they are doing. More and more often, deception, exaggerations, or vague references to an informer are used to get the warrants in the first place. These people had absolutely no probable cause in this case, but that doesn't seem to matter anymore. Does this sound more like the Soviet Union than the U.S.? Stay with me; you haven't heard anything yet.

HARRY LAMPUGH. Harry was sitting at his kitchen table in just his pajamas. His wife, Terry, was in the bathtub when she saw someone peeping through the bathroom window at her in the tub. Suddenly, the front door burst open without warning and more than a dozen BATF and IRS agents poured into the house with guns drawn. No one had on a uniform. No one offered any identification. They did not announce who they were or why they were there. No search warrant was presented.

When Harry asked if they had a search warrant, one agent stuck his gun into Harry's face and said, "Shut up". The agents then spent the next three hours tearing the house apart. They opened safes, cabinets and drawers. Furniture was turned over and smashed. Papers were scattered all over. The Lamplugh's three cats were killed. (One of the agents stomped a prized registered kitten to death in front of them.) The agents refused to let them get dressed and accompanied them to the bathroom the entire day. They were held prisoners in their own home at gunpoint, all day, without any idea of what these people wanted.

During the raid, the agents decided to have lunch. A couple of agents went out to get pizzas, and then they sat down to have a party. Acting like juvenile delinquents, they threw half-empty soda cans, pizza, and wrappings around the house. Then, they went back to trashing the house for another three hours. They emptied about twenty bottles of Harry's medicine that he takes, mostly for cancer, all over the floor. The only clue that Harry has as to what they were there for is that one of the agents asked if he had a machine gun. Of course, he didn't. Eventually, he found out that they were *aparently* talking about a commemorative gold inlaid gun which was a collector's item; it was a perfectly legal semi-automatic, though Harry hadn't had it for years. When he described the gun to the agents, one of them said, "That must be what *they* were talking about." *They Who?* Harry still doesn't know.

Was completely trashing their house, killing their cats, and terrorizing them for a day enough? Oh, no. The agents then took almost all of their records, including marriage and birth certificates, insurance papers, school records, vehicle titles and registrations, and his medical records. None of these items were mentioned in the search warrant, and they had no legal right to do anything with them. Can anyone in their right mind justify taking these personal items, even if Harry would have had an illegal gun? This is nothing but deliberate, illegal harassment. Despite the fact that the Lamplughs had committed no crime, were not even suspected of one, and nothing illegal was found in the search, the agents still illegally confiscated (stole) over sixty perfectly legal guns and all of his ammunition, valued at about $70,000.

They also took all of Harry's large lists of newspaper contacts, and all friends' and family's phone numbers. They opened his mail and read it all, in direct violation of U.S. postal laws. Can you think of any possible justification for this nonsense? Next, they carted off all of his records of over 70,000 names and addresses of gun show contacts and exhibitors.

Ah, ha! Now you know why they were there. Harry was on their search and destroy hit list. Harry helped organize gun shows in the Northeastern U.S. What he does is completely legal, but the BATF vendetta against gun shows knows no bounds. They took all his records in order to put him out of business – a perfectly legal business. This whole mission was nothing more than a bunch of terrorists with badges trying to make sure Harry was never involved in any more gun shows, and also to send a message to others so involved that the same thing could (would?) happen to them.

The records illegally seized in this raid and others are being combined with the other records the BATF illegally obtains in direct violation of the law by intimidating gun dealers and pawn shop owners into letting them copy their records. This is going on at their new firearms tracing center in West Virginia. This is de-facto gun registration. Congress, where are you?

What about the warrant they supposedly had, but never showed? It turns out that it did not name even one specific item. According to their attorney, such vague warrants are unconstitutional. So what? Most of the warrants obtained today are, even when these agencies bother to get one. An obvious question is, what is Harry supposed to have done to provide any legitimate grounds for even issuing a warrant? No one knows. The affidavit for the warrant has been sealed. Why? No one can find out.

Other agents invaded the home of Harry's son John. Harry's youngest son was there, home from college. The agents confiscated $2,000 of his money. When he asked for the money back, they laughed and said he might get it back someday if he could prove it was legally his. Back at Harry's house, one agent found a few dollars and a grocery list in the pocket of Terry's coat. Grinning, he said she could have the money back only if she could give him the serial numbers of the bills from memory. Harry had ordered a new truck that was to be delivered that morning. The money and the paperwork were on the dresser, waiting for the truck. Both were confiscated. Next, they found over a thousand dollars in a drawer that Terry was saving up for some needed surgery. When she complained about them seizing her money, an IRS agent said, "We'll see how cooperative you are when we throw you in a cell full of lesbians." These people *must* be shut down or it will only get worse and we will *all* be at risk. If they can trample one person's rights with impunity (actually thousands at this point), they can do the same to you.

> *"It is dangerous to be right when the government is wrong."* - Voltaire

The Lamplughs are trying to piece their lives and their business back together and make enough money to continue their legal battle with the BATF. At least four grand jury witnesses have said that the statements read to the grand jury were not the statements they gave to the BATF. These were deliberate fabrications to try to get an indictment to cover the BATF crimes. Another witness said he was threatened with charges of perjury if he didn't change his testimony to what the BATF wanted. He refused. The Lamplughs have been repeatedly threatened by "anonymous" phone calls. Terry Lamplugh has been repeatedly stalked by BATF agents. One time as she approached her van, one of the men who had been stalking her suddenly appeared at her side and said, "Don't ____ with the bureau." As he walked away from her he called back, "Check out what's in your van." Inside she found a dead black cat with a broken neck, a reminder of what they had done to her cats. This is in America, folks. These are federal law enforcement agents, sworn to uphold the law, paid with your tax dollars to protect you from this sort of thing. How do we tell the crooks from the good guys anymore?

These agents committed about a dozen felony crimes, and many misdemeanors. The local county sheriff should arrest everyone of them, for everything from cruelty to animals to assault with a deadly weapon. There is a bill pending in Congress to require the permission of the local sheriff before these federal agencies can conduct a raid. That is a good start, but the real question is, Why should they be allowed to conduct any "raids" at all? The founding fathers were deathly afraid of a national police force, for good reason as we are now seeing. Law enforcement was to be in the hands of local sheriffs, responsible to and for their communities. The FBI is (was?) the Federal Bureau of *Investigation*, not enforcement. At first, they weren't even allowed to carry guns. Now they have trained snipers and para-military assault teams whose job it is to kill people. The bill in Congress does not go far enough. At least, it should require that at least two deputy sheriffs accompany any such raids, not as participants but as observers. They should be empowered to intervene whenever they feel that the objects of the raid are having their rights violated, that items not clearly specified in the warrant are being taken, or that any law is being broken.

Even without the bill, your local sheriff can be your best protection against these kinds of federal overkill. If you don't have one that will stand up to the federal agencies, elect one, regardless of what party he belongs to. If local sheriffs will start arresting these thugs and local prosecutors will prosecute them, some of this terrorism would stop, at least where sheriffs and prosecutors are elected who would stand up to them. Of course, if the local sheriff is getting a cut of the spoils under the confiscation laws, then you have a serious problem. Congress should bring these kinds of activities to a halt, but even with the Republicans in charge, they are going the wrong way. The courts could put a stop to some of it by requiring probable cause or real evidence of a crime before issuing a warrant, making sure that the warrant is very specific. However, since most judges are more liberal than the law enforcement people, and since they also get a cut of the spoils confiscated in many cases, don't hold your breath for any relief from that quarter. If the promotion-driven-glory hounds at the Houston BATF would have gone through the local sheriff, Waco would never have happened.

THE REAL WACO WACKOS.

The liberal media did a superb job of smearing David Koresh. They even managed to compare him to Jim Jones on one extreme and fundamentalist Bible-believing Christians on the other. And for the most part, people bought it.

The people who know the facts have a far different picture of what happened in Waco. The whole event started out as a staged show for the media by the BATF, to make a point to anyone who might not want to go along with the government's growing elimination of personal and religious freedom. It was also the bright idea of the Houston office that this attack would somehow enhance the badly tarnished image of the BATF. The supposed excuse of thinking that "maybe" the group was converting legal semi-automatic weapons to fully automatic capabilities was utterly transparent to anyone familiar with the circumstances. The BATF had been labeled by Congress as a renegade group that was completely out of control long before Waco, but Congress has done nothing to rein them in or hold anyone responsible for their many violations of our most fundamental rights under the Constitution.

Months before the assault, the McClennon County Sheriff had called Koresh and asked if they could come out and inspect their weapons and papers. They were *invited* to come in. They thoroughly checked every weapon and every bit of paperwork. All were in perfect order. The BATF knew all of this.

On July 1992, two men showed up at the home of Henry McMahon, who was a gun dealer in Hewitt, Texas. One was neatly dressed in a business suit. The other was wearing blue jeans and a t-shirt. When they identified themselves as BATF agents, they were asked for identification. The one in the suit, Jimmy Skinner, handed over his identification card. The other agent, Davy Aguilera, had no identification of any kind. Mr. Skinner said that Mr. Aguilera was a trainee which was a lie. Mr. Aguilera had been an agent for five years and was a major instigator of the Waco disaster. Seven months later, Aguilera made up the affidavit that was used to obtain the warrant to try to justify the assault on the compound.

While the agents were in McMahon's home examining the records, they began asking questions about David Koresh. Mr. McMahon called Koresh and told him BATF agents were there and asking questions about him. Koresh responded by saying, "If there is a problem, tell them to come out here. If they want to see the guns, they are *more than welcome.*" McMahon then walked into the room where the agents were, carrying his cordless phone. He told the agents that he had Koresh on the phone, and that if they would like to go out there and see the guns, that they were more than welcome. Aguilera said, "No, no." They were invited to the compound as every other official had been, and they refused to go.

It is now known that the BATF was planning an all-out assault on the compound at least seven months before it took place. Practice assaults were held in several different places including a vacant farmhouse about five hundred feet from the Mag Bag, where the Koresh group restored cars as one of the businesses that supported the them.

Both McMahon and Koresh kept meticulous records. McMahon notes that Koresh was a real stickler for crossing every "t" and dotting every "i". He wanted everything to be perfectly legal so that no one could falsely accuse him or the group of any wrong doing. Koresh was very particular about getting the original boxes the guns had come in. When asked why he wanted everything original, he replied that gun collectors want everything to be original if possible and that the group would make more money when the guns were resold. When asked why he bought so many guns, Koresh responded that with the government heading toward banning more and more guns, the prices would go up and they could make a profit to support the group. He was right. Prices have soared on many of the types of guns he purchased. As an aside, the treatment of McMahon and his business partner, Karen Kilpatrick, by the BATF after the initial raid reads like something out of a B-grade movie about the KGB. The BATF was determined to keep them from talking to the FBI and the Texas rangers, who had started investigating the BATF, about what the pair knew about the BATF's actions. Later, the FBI also became embroiled in the whole mess, and has gone to great lengths to cover up its own conduct, as it always has done in the past. These two people never did anything but try to cooperate with the various law enforcement agencies, and yet their lives have been shattered. The BATF even got McMahon fired from his job. If they haven't filed a multi-million dollar lawsuit against the government, they should.

Koresh had always granted free access to any law enforcement people, social workers or other agencies who asked. They never needed a warrant. Why didn't the BATF just call and ask like the others before them? They already had been invited in months before. If all they wanted to do was serve a search warrant, why did they take a massive assault team of over a hundred people and invite the media to watch? If you don't think this was a staged media event, why was it named "Operation Showcase"? They made no attempt to serve the warrant. It was fraudulently obtained to start with and was just a cover up for the planned assault. David Koresh had committed no crimes. If they wanted to arrest him for something, why didn't they do so when he was outside the compound, which he often was? He asked that question after the initial raid and they (and the media) ignored the question. Congress should ask this agency the same question. Why didn't they just ask him to come in? He had before whenever asked. They didn't want Koresh; they wanted to stage a showcase assault for the media.

The sheriff had been to the compound several times, at Koresh's invitation, when the sheriff had called to check allegations of weapons violations. One time he took every single gun to his office to thoroughly check everything against the paperwork. Not even the slightest irregularity was found, and the guns were returned. The BATF knew this. According to the sheriff, the Davidians had fewer guns per person than the average Texas resident.

The BATF staged a violent massive assault on people who were minding their own business and bothering no one, simply because they had bought a number of guns and believed in the right to defend themselves if attacked. They did have that right, both under Texas law and common law that has been recognized for thousands of years. Unfortunately, they had been targeted as one of those "right-wing fundamentalist Christian extremist groups" that these agencies seem so paranoid about and against whom they seem to have an irrational vendetta. Apparently, our government seems to forget that the pilgrims and others who settled this country were the fundamentalist Christian "cults" of their day, who simply wanted to be left alone to follow their beliefs and had to flee from government oppression in order to do so. Is history repeating itself?

On February 28, 1993, the last day the fraudulently obtained warrant was valid, two pickup trucks pulling cattle trailers sped down the road to the compound. As the men in the trailers jumped out, they ran screaming toward the compound like some cavalry charge up San Juan Hill. They were *not* identifying themselves as law enforcement agents or anything else. David Koresh saw them through a window and stepped out onto the front porch, unarmed. According to several witnesses, he held up his hand and said: "Wait. Go back. There are women and children in here. Let's talk about this." At this point, one of the charging agents fired at him. The bullet barely missed his head. As he dove through the door and closed it, a hail of bullets hit the door and also the windows on the second floor where a number of the women were looking out the windows. Koresh was wounded before he could get the door closed. A man standing behind him was mortally wounded. Neither Koresh nor anyone else inside was armed at this point. Jack Zimmerman, a Houston lawyer who represented Steve Schneider, says that the BATF version of what happened was a lie. He saw the front door that was peppered with bullet holes. The BATF also said that they only fired at identifiable targets. They lied. There were no identifiable targets behind those doors or in many of the rooms they fired into indiscriminately.

According to BATF agent Roland Ballesteros, the BATF shot first and made no announcement that they were federal agents. Agent Eric Ever confirmed that the first shots he heard came from a team of agents with dogs.

The first notice that Steve Schneider had that something was wrong was when bullets from the three helicopters flying overhead started coming through the ceiling. Jaydean Wendel was lying in bed just after nursing her eleven-month old baby when a bullet from a helicopter went through her head. Another man was killed by the bullets from the helicopters while sitting on a bed. These people didn't even have time to move. They probably never knew they were under attack. Within a minute of the initial attack, Wayne Martin placed a frantic call to 911 pleading with them to call off the assault. On the tapes, you can hear Steve Schnieder's voice in the background saying, "Here come the helicopters again." Peter Gent was working inside the water tower. When he heard the commotion, he stuck his head out to see what was going on. He was shot through the head by a BATF sniper. None of these people were armed. It appears that nobody inside was armed at the time. The team leader of the sniper team has stated that they knew that at least three of the people they deliberately killed were unarmed. These agents violated virtually every standing procedure, including not firing indiscriminately into buildings especially those containing innocent people or children.

Mike Schroeder was a member of the Branch Davidians who arrived at the compound after the BATF had set up a perimeter. The Texas Rangers stopped him and two others and searched them. Then the rangers let them go on into the compound. Before they could reach the building, they were ambushed by the BATF. Mike was shot seven times. Most of the shots hit him in the back. This was cold blooded murder. The other two managed to get into the building. His body was left to rot for several days. Later, the ground around him where he was shot was removed to a depth of a foot. What was the BATF covering up?

Many of the group's guns had been packed to take to a gun show for sale or trade, which is one way they earned money. All of these guns were perfectly legal and Koresh had all the proper receipts and paperwork. Other guns were still in their original boxes. When they opened the boxes to get something to defend themselves with, they discovered that there were no magazines or ammunition. This was hardly an armed militia lying in wait for government agents. Video tapes and witnesses confirm that the people in the compound were under very heavy attack with automatic weapons from the ground and the helicopters for at least ten minutes before firing back.

Under Texas law, it is perfectly legal for people to defend themselves, even against law enforcement officers who start shooting, especially if they haven't identified themselves. Even when the group started returning the fire to keep out the people who were firing into the house as they came in, they still had not committed any crime. They were not even resisting arrest as no attempt (or announcement) had been made to arrest them. They still had not been told who these masked men were.

Videotapes also show three agents going through a window. Then, an agent outside the window started firing an automatic weapon into the room through the window. The three agents died, almost certainly from this agent's fire.

The FBI kept telling the American people that they were urging the people in the compound to surrender. Yet on April 17, a bureau spokesman announced that anyone who came out would be shot. That same day, when one individual tried to leave through a window, he was driven back by gunfire and grenades.

The door with the bullet holes has conveniently disappeared. It is not in the pit where the rest of the compound structure – what was left of it – was quickly buried. The roof and ceilings that were penetrated by the shots from overhead were conveniently burned up and anything left was buried in the pit. The government made sure there was no physical evidence left of what they had done. Local officials were not allowed free access to the site before it was bulldozed so that they could investigate the evidence.

The attack helicopters were illegally obtained from the Texas National Guard. By law, they can only be used in a raid where drug enforcement is the principle reason. In the request for the helicopters, the BATF lied on the request saying they were to be used in a drug raid. This assault never had anything to do with drugs, which is not the province of the BATF in the first place.

The FBI put out the story that the group committed mass suicide when the final assault began. No way! The whole suicide fairy tale was concocted several weeks in advance of the final assault and played up in the media as a cover story for the "final solution". Apparently, they were preparing the public for the outcome that there were to be no survivors. Koresh was planning on writing a book on the whole affair, which could not be allowed to happen. An FBI sniper has testified that he saw a masked figure dressed in black spreading liquid around on the floor during the final assault. These were the uniforms worn by the FBI's hostage and rescue teams. When the sniper's remarks became public, the media put out a story that the Davidians had been buying black cloth and sewing black uniforms. They were not. Why would any of the Davidians be wearing a black uniform and a mask inside their own compound? The FBI sniper has been silenced. In a lawsuit filed by former U.S Attorney General Ramsey Clark, it is charged that the FBI put several agents into the compound as the final assault began. They either snuck in the back door where the media couldn't see them from so far away, or they were injected by one of the CEVs (Combat Engineering Vehicles) that were battering the walls down. They were in there about five minutes.

The FBI claimed that Koresh and others committed suicide. Clark claims that the FBI team murdered Koresh and the other leaders. Koresh was killed by a high-velocity bullet that entered his forehead just above his left eyebrow and went out his right ear. Koresh was right handed and his left hand was disabled by a wound to the wrist. He could not possibly have shot himself that way. What about the other leaders? They were *executed* by bullets through the back of their heads. By whom? Once the fires started, everybody was desperately trying to get out of the building.

The well-planned and executed CVE maneuvers took out the staircases, so those upstairs had no way to escape the flames. One of the drivers of the CVEs said it was his assignment to crash through the walls and bury the entrance to the underground storage area, where the people inside the building would have been safe from the fire if they could have gotten to it. There were to be no survivors. The only reason there are any survivors is that the few people who did manage to get out were in plain view of the media, and the FBI could hardly mow them down, as they apparently did to some who escaped out the back where the media could not see them.

The BATF had been practicing the assault on the compound for almost a year. They had been studying drawings of the buildings for a long time. Before the final assault, the former McClennon County District Attorney predicted that the government would kill the entire group. He later stated that they intended to kill these people from the beginning to cover their own unlawful activity. When he was the district attorney, he had thoroughly investigated Koresh and the group and found nothing wrong.

Thirty-two of the bodies were found in the central storage bunker. They had wet blankets over their heads trying desperately to survive the fire. Some suicide pact. What it is is a pact (or pack, if you prefer) of lies by law enforcement officials and government officials at the highest levels. The central storage area known as the bunker, was actually a good place to try to survive the fire. It was a thick concrete structure with steel doors. It had survived a previous fire and would have survived this one. Clark's suit alleges that the fireball that erupted during the inferno was not a propane tank as the FBI claimed, but an incendiary bomb used to make sure everyone in the bunker was dead.

Why did Janet Reno approve the use of CS gas? It is not designed to *drive* people out, as is CN tear gas. It is designed to completely *immobilize* people so they *cannot* get out. It is so dangerous and lethal that it is outlawed in war. We have signed an international treaty to that effect. It cannot be used, even in war, without the express written approval of the President. It is particularly lethal against children. It is never supposed to be used in enclosed places. Furthermore, the propellant used was highly explosive. CS gas is about ten times more flammable than natural gas. Why did they use it if they didn't intend to burn the building down?

Normally, harmless carbon dioxide is used as a propellant for CS to reduce its explosive flammability. The FBI claimed that this is what they used. They lied. Instead, they used acetone and ethanol, which are extremely flammable. These chemicals are known as great accelerants for starting very hot, fast spreading fires. This is probably why the flame fronts spread so rapidly through the compound. Significant amounts of both acetone and ethanol were found in the lungs of the victims, so there can be no denial of what they used. Not only is CS highly flammable, but it breaks down into deadly cyanide gas at high temperatures, as in a fire. There were to be no survivors, just as the former district attorney stated. If the President authorized the use of CS, he was lying through his teeth when he said it was just an irritant-type tear gas on national TV. If he did not authorize it, who broke the law by authorizing its use, and again, *why*?

The FBI knew the place was a tinderbox with wood walls, hay bales, and gas or kerosene lanterns everywhere, because the electricity had been cut off. Then they filled the building with highly flammable CS and propellants to make sure the place virtually exploded when the assault team started the fires. Experts on the use of CS have stated that the amount used was incredibly excessive and far beyond what was needed to immobilize the people inside. Obviously, they had other reasons than just immobilizing them.

An FBI infrared aerial photo clearly shows that the first fire was started in the gym behind one of the CEVs (Combat Engineering Vehicles) that had demolished the gym. A Justice Department report confirms that their people were in the gym. Yet, the FBI claims that the first fire started about ten minutes later in a different part of the building. A second fire was started by a CEV knocking over a lantern. Of the three probable ignition points for the fire identified by the FBI, they were in control of at least two of them at the time the fires started. The Branch Davidians did not start the fires.

The FBI knew about the virtual certainty of a fire. It is clearly spelled out in the CS manual. Why, then, were the fire trucks that had been standing by for seven days sent away just before the assault, unless they planned to burn the place down? More damning, why was the fire department not called back until long after the assault team started the fires and the building was fully engulfed? Even then, they were not allowed to fight the fire for another hour until the buildings had burned to the ground and everyone inside killed. Why? Could it possibly be that the FBI wanted to make sure any evidence of their actions was destroyed? At first, the FBI denied that they blocked the fire trucks from coming in. Later, after comments on TV by fire officials, they changed their story and said they kept the fire fighters away because of fear of bullets from the people in the building. Sure. What bullets? These people were desperately trying to get out or find a place of safety if they couldn't get out. The holes punched in the building and the forty-mile-per-hour winds did sweep away enough of the gas that some of the people were able to recover enough to try to crawl to places of protection, to no avail. Video tapes clearly show FBI personnel standing around close to the building watching the fire with no vests or protective gear. They certainly were not afraid of any weapons fire.

This is just the beginning of the unanswered questions. Why were the remains of the compound bulldozed into the basement and all the evidence covered with dirt before any local authorities, even the Texas Rangers, could investigate anything? The FBI's "independent expert" who said the Davidians started the fire, is a former BATF employee and a contractor for the FBI. His wife is the personal secretary to the head of the Houston BATF who planned the assault. Some independent investigator! It really strains credibility to believe that the media is so incompetent that they gave so much coverage to this so-called expert without even checking out who he was.

Why did the FBI agents in charge reject the advice of the FBI psychological team that said, early on, that escalating the pressure was exactly the wrong course of action and would only harden the resistance? The team developed a plan that would lead to the easy resolution of the problem. They were ignored. Why? One explanation might be a statement by one of the agents that these Davidians had thumbed their noses at the government and had to be made an example.

If the full truth is ever brought out, the Waco cover-up of the deliberate massacre of the Branch Davidians, who had become an unacceptable embarrassment to the government agencies involved, will make Watergate look like a Sunday school picnic. Will Congress ever really investigate what happened there? Perhaps now, after Oklahoma, if enough people insist on it. Could Oklahoma have been avoided by a real congressional investigation in the first place? We will never know. In Clinton's administration, the foxes are in charge of the hen house. The little "interview" that the democratic Congress conducted was a sad joke. The internal investigations were even sillier. Two BATF raid commanders, Phil Chojnacki and Chuck Sarabyn were fired, not for the disaster they created, but for lying and altering documents to divert blame to someone else (typical bureaucratic response). The reason given for their firing is irrelevant. They were later rehired, apparently after they threatened to tell the truth about Waco. There are some high-level people in the government agencies involved who should be tried for murder and conspiracy. Will they be? Not by this administration. Twenty fully documented lies have already been put forth by these people. One thing that does puzzle me is how the Texas Rangers have been effectively silenced. Normally, these people are not afraid of anyone, and this attack was against their citizens.

In case you are interested, one of the reasons the BATF attacked Koresh, according to their affidavit, was that he had been critical of the BATF and owned a video put out by the Gun Owners of America that was critical of the BATF. Have you criticized the BATF lately?

Despite the severe threats against agents if they talk, some cracks are starting to show in the cover up. Letters from disgusted agents are showing up in *The Agent* which is the official publication of the National Association of Treasury Agents. One letter denounced the use of inexperienced agents as cannon fodder at Waco. Maybe using these inexperienced agents is the cause of the agents being killed by friendly fire. Another letter said experienced agents would never have followed Chuck Sarabyn's crazy orders. Still another letter compared the BATF commanders to those in his Vietnam experiences, who were so promotion minded that they would willingly sacrifice the lives of their men for a little glory. If these letters are any indication, maybe there is hope for the BATF. I doubt it. This agency needs to be disbanded for several reasons. One is that it is so shot through with the assault team mentality and willingness to trample on people's rights that it is probably irredeemable. Another very important reason is to send a loud and clear message to other agencies, in the only way they seem to understand, that this kind of behavior will not be tolerated by the American people.

Waco was a publicity stunt to show how the government was cracking down on cults with "illegal" arms caches. Waco was the first big test of using federal agents, trained by the military, to use military assault tactics. It was a test of using the media, which was kept at a distance where they couldn't really see what was going on, to broadcast the government's propaganda. The media dutifully reported what it was told. The public was deceived. Government agents murdered all those innocent people right in front of the cameras and convinced the public that the Davidians had brought it all on themselves, even committing suicide.

Is anybody interested in a real congressional investigation? The second investigation, this one by the Republicans, will do little to clear the air. Government witnesses and "experts" lied through their teeth. Call your congresspeople. You have permission to copy all or part of this chapter providing you indicate the source, for the purpose of sending it to your congressman. Demand answers to your questions. Don't let them get away with form letters for answers or the excuse that they have already investigated these things. Stay on their case. Let's have the whole truth.

Despite the Waco disaster, the BATF continues to plan military assaults on innocent citizens to show off for the media in order to polish their badly tarnished image. Apparently, they still think that one big successful showcase raid, like Waco was supposed to be, will get the public back on their side. The latest such massive military operation was to be against the Kelly Miller public housing complex with more than 200 agents. It got called off at the last moment when the BATF learned that a press release with the details of the raid was issued prematurely.

Besides all of the raids and killings, the BATF continues their campaign of harassment against gun dealers. The goal is to shut down or intimidate as many Federal Firearms License dealers into leaving the business as possible, in order to dry up the outlets for firearms and ammunition purchases. One of their targets was the dealers in Pueblo, Colorado. Due to their illegal harassment, 65 of the 125 dealers "voluntarily" gave up their licenses and went out of business. In Colorado Springs, 105 of 126 dealers gave up their licenses. Denver and New York City are their current targets. *This is illegal*. Congress, are you going to rein in the dogs you set loose?

FOOD AND DRUG ADMINISTRATION

The quote by the current head of the Food and Drug Administration at the beginning of this chapter should give you a pretty good idea of the mentality of this organization. The FDA is supposed to protect the health of U.S. citizens. Yet, this agency is probably responsible for more American deaths *each year*, directly or indirectly, than there were in the Korean and Vietnam wars combined. Is this a wild, irresponsible statement? Well, you make up your own mind after you take a look at some facts.

On July 29, 1993 at a House subcommittee meeting, David Kessler presented a report entitled:, *"Unsubstantiated Claims and Documented Health Hazards in the Dietary Supplement Market."* His comments were dutifully reported in the national media as gospel truth.

After the meeting was over, Senator Orrin Hatch, the co-sponsor of legislation to protect health-care freedom in the U.S., asked his staff to analyze the report. Their report was issued October 21, 1993 and was never mentioned in the media. One of several quotes from the report is, the report "should be immediately withdrawn and the FDA should apologize to the Congress and the public for its release ....The FDA has knowingly submitted false information to Congress." Among all the charges against the FDA in the report are some interesting tidbits. One of them is that the FDA willfully stated that a paperback book is a nutrient product. Another is willfully listing so called "unsubstantiated claims" for products that don't even exist. Still another is intentionally misleading the media and defaming innocent nutrient supplement companies. The list just goes on and on. How about intentionally spending $25,000 (of your tax money) in an attempt to make Dr. Kessler look good before Congress and the media?

From the *Life Extension Report*, February 1994: "The most deceptive act was to label ALL claims for products made by health-food employees 'unsubstantiated' without ever asking for them to be substantiated. Having the FDA be judge and jury for nutrient supplement claims is outrageous because the FDA knows little or nothing about nutrient supplements, because the FDA has been biased against nutrient supplements since its inception and because the FDA employs illegal and unconstitutional tactics in its attempt to exercise totalitarian control over health-care in this country."

Totalitarian? Hmmm. It sure sounds like Kessler's statement at the top of this chapter could be understood that way. The FDA's fanatical drive to eliminate anything that they have not approved of in one of their hopelessly expensive procedures (about 200 million dollars to get a drug approved) is well known by anyone associated with any form of alternative health care or the health food industry.

Bureaucracies Gone Berserk - 45

Their methods are often illegal and frequently so silly as to defy explanation.

Before we get to the flagrant abuse of our rights and the many illegal acts of the FDA, another indication of their attitude in general is their intention to build a monument to their own egos. While the government is running ruinous deficits and supposedly trying to rein in spending, the FDA wants to build their own version of the Taj Mahal. The statement below is from a speech by Representative John Duncan in the House of Representatives on January 31, 1995. If you are fed up with struggling under your current tax load to support government waste, you might want to take an FDA-approved tranquilizer before you read it. On second thought, the side effects from the tranquilizer may cause you to do something rash, so scratch that idea. Instead, you might try some harmless relaxation techniques that the FDA is trying to outlaw.

"Mr. Speaker, one of the most wasteful, inefficient agencies in the entire Federal Government is the Food and Drug Administration (FDA). In their bureaucratic and arrogant way, they have held medicines and medical devices off of the U.S. Market, sometimes for years, to the detriment of the health of American Citizens. (They died!-RT) *By their rules, regulations, and red tape, they have driven up the price of drugs and have helped the big drug giants* (with whom they have a fat job waiting-RT) *by making it extremely difficult or almost impossible for small businesses to compete in the field.*

"Now, however, they want to do something which should outrage every taxpayer in the nation. At a time when we are supposed to be downsizing the Federal Government, the FDA wants to build a Taj Mahal complex of buildings in Maryland for a new headquarters. Part of this project is to be in Montgomery County and part in Prince George County.

"However, the important point is not the location, but the cost. The original cost estimate for these buildings was almost one billion dollars. However, because the FDA has become concerned about the appearance (the typical bureaucrat's only real public concern-RT) *of this exorbitant and excessive spending at a time when most people want frugality in Government, they have lowered their estimated cost, all the way down to $875 million. Even if this project*

comes in on budget, which I seriously doubt, it would still be at a cost of whopping $257 per square foot. State governments are building beautiful buildings for half this cost.

"And is the FDA doing everything possible to hold down costs? Well, since the money is not coming out of their own pockets, they chose the most expensive site they looked at and some of the most expensive land in this Nation. The original estimate of the Price George facility was $290 a square foot. The Montgomery County complex is to be several buildings interconnected, in a college campus-like setting, on a 530 acre tract of land – 530 acres when they could build a beautiful headquarters on an acre or less.

"The fact is, Mr. Speaker, the FDA should be greatly reformed. It should be greatly downsized. It should stay where it is now. (It should be abolished-RT)

"Perhaps the most phenomenal thing of all is the size of this project – 3.4 million square feet – to house only 6,500 employees. This comes out to approximately 750 square feet per employee. (The size of a small house-RT). *Most members of Congress have approximately one-seventh of what the FDA wants.*

"With a national debt of more than $4.7 trillion, we should not be spending almost $1 billion to build plush new quarters for FDA bureaucrats. The bureaucrats want to live like kings while the taxpayers foot the bill (and control every aspect of the serfs' lives-RT).

"I know that we are a government that is of, by, and for the bureaucrats instead of one that is of, by and for the people, but, Mr. Speaker, this is one I hope that we can win for the taxpayers."

As I previously mentioned, Mr. Duncan also referred to how detrimental the FDA has been to our health. There is a drug that is widely used around the world and is badly needed in this country. Why has the FDA not approved the use of this drug? Isn't it safe? Yes, it is a very safe drug. Isn't it effective? Very effective. Then why on earth is the FDA refusing to approve the drug? Answer: the FDA is punishing the company for irritating somebody at the FDA, apparently by advertising the drug before final approval. They are making an object lesson out of them and trying to drive them out of business in a ridiculous petty vendetta. In the process, we are denied access to a badly needed drug.

What was this heinous crime that should so infuriate the bureaucrats? Well, it seems that the company broke the FDA's unspoken rules that you can't advertise a drug before it is approved.

If that sounds like a petulant ten year old, you haven't heard anything yet. The even greater danger is that these people have badges and guns at their disposal, and virtually a blank check when it comes to making their own laws and regulations. They have shown no reluctance to abuse their powers to attack and eliminate anyone that doesn't line up with their particular view of things. They have shown even less reluctance to assume powers Congress never gave them or intended them to have. Constitutional rights? Forget them.

The FDA has become the enforcement agency for an industry they are supposed to be regulating. The attempted power grabs of the FDA dwarf anything the other agencies are doing, which is bad enough. They are trying to reclassify almost everything you put in your mouth as a drug so they can regulate and possibly ban it. In the FDA's case against Nutri-Cologys, the judge stated; "It is apparently the FDA's view that if a company makes a claim that milk helps prevent rickets, milk suddenly becomes a drug." It even gets sillier than that. If you add water to a food product, it becomes "adulterated" and can be seized by them. They are trying to reclassify recreational and relaxation products as medical devices so they can regulate those as well. Their greed for power and expansion seems to know no bounds. Here are just a few of the possible thousands of examples from the Federal Bureau of Ridiculous and Illegal Raids (FDA).

Synetic Systems is a company that makes *entertainment* and relaxation products that contain gentle flickering lights and soft gentle music or soothing sounds like those of a running stream. In the summer of 1995, the FDA raided Synetic and seized all their inventory, leaving nothing for them to ship to their customers. Did Synetic violate any law? No. Did they violate any FDA regulation? No. Did they do anything wrong? No. Well, what on earth did they do to cause this raid? Who knows?

The FDA *claims* (which is their usual tactic for expanding their power) that they have jurisdiction over Synetic's products because they are somewhat similar to things used by some neurologists to *affect* brain waves. *What*, pray tell, *doesn't* affect brain waves? As the President of Synetic points out, the FDA's contention that his products are "medical devices" because they might affect brainwaves is ridiculous, because everything you do or think affects your brainwave activity, from reading a book to watching TV, listing to music, or making love. Apparently, the FDA thinks it has jurisdiction over anything that affects our minds, which is nearly everything we say or do. What's next – motivational courses, television, music CDs, movies, or books? Don't laugh; their next target is relaxation tapes on cassettes. After that, they are going after exercise machines as medical devices. Even that makes more sense then the Synetic raid. Maybe we could get them to shut down the discotheques with deafening music that causes irreversible hearing loss and blinding flashing lights that make you dizzy. No, there is little chance of them going after anything that actually causes harm.

What follows next is a classical case of FDA lunacy. How does Synetic get permission to sell its products? The FDA insists that they *prove* that they are safe and effective. *Safe*? Soft music? Soft lights? Come on folks, get real. About the only danger from soft lights and soft music is that somebody might end up doing something that results in pregnancy. Are they going to outlaw that? *Effective*? Effective at doing what? The FDA doesn't say. Synetic makes no medical claims for their products to be effective for any medical anything. In other words, there is *no way* to get their products approved. They asked the FDA how they were supposed to do this. The FDA won't tell them, which is another classic FDA tactic. In other words, there are no regulations, no guidelines, no definitions, no procedures, and *no authority* for the FDA to do this, but they just do it anyway. Wouldn't it be nice if the FDA required the same "safe and effective" standards for the drugs they unleash on us from their buddies in the big drug companies?

I am going to take a slight detour here before going back to the illegal raids. On the subject of safe and effective drugs, the FDA has approved a group of drugs for regulating heart arrhythmia. The FDA had irrefutable evidence that these drugs were killing people by *causing* abnormal heart beats when they approved them. Their excuse was that they *thought* that more people would be saved by the drugs than *killed* by the drugs. Did they have even an ounce of evidence that this theory was true? No! They never required the drugs to go through controlled clinical trials to even prove they worked.

When the National Heart, Lung and Blood Institute conducted large-scale tests on three of these drugs, they discovered hard evidence that these drugs were killing people in large numbers. Did the FDA withdraw them? No, they were too busy raiding Synetic. Synetic must somehow prove that their soft lights and soft music are safe and effective, but there is no such requirement for these drugs that are killing people. Could it be that the FDA is going after anything in the "natural" area that might somehow compete with the cut and/or drug medical establishment that is so profitable to the current medical power centers?

The public outcry over the findings about these drugs caused the FDA to investigate themselves. Right! Who did they appoint to do the investigation? Why, the same man who had approved the drugs in the first place, of course. So much for internal investigation. Congress, where are you? The outcome of this "investigation" was to recommend some changes in the labeling on the medicine. Peachy. If you want more gruesome details on this deadly fiasco, get the book *Deadly Medicine* by Thomas J. Moore. If your doctor wants you to take any of these medications, get a copy of the book and insist he read it first.

The FDA has become a product promotion agency for the industry they are supposed to regulate. They are on a search-and-destroy mission against any kind of alternative medicine, except chiropractic medicine which has too much political clout. They particularly go after anything that has to do with nutrition, vitamins, and supplements. I could write a book about all of the raids and illegal activities, but if you want all the gruesome details, you can get more information from the Life Extension Foundation. If you are really interested in this subject, you might want to visit their FDA Holocaust Museum. The address is in the Appendix.

In 1987, the FDA raided the Life Extension Foundation. Eighty percent of the things they took to try to put the foundation out of business were not in the search warrant. The FDA agents involved testified under oath that it is *official FDA policy* for all such raids to take whatever they want to take *regardless* of what is in the search warrant. This is becoming standard procedure for most federal agencies even though it is completely unconstitutional. They get a search warrant on whatever flimsy excuse they can think of and go on a fishing or punitive expedition without serving the warrant or even paying any attention to what is in it.

What was the foundation's crime? They reprinted articles or abstracts from respected medical journals about drugs and treatments that the FDA had not yet approved in some cases. Since only the FDA can determine what is "truth" and what is good for you, the Life Extension Foundation was a direct threat to them. Results: armed raid and confiscation.

Dr. Jonathan Wright, M.D., is a prolific writer on topics concerning nutrition and supplements. He is therefore a real thorn in the sides of the FDA and the medical establishment. On May 6, 1992, the FDA with six armed deputy sheriffs, raided his office, seizing vitamins, allergy screening equipment, computers, bank records, mailing lists, and patients' medical records to try to put him out of business. No charges have been filed.

A young woman in Phoenix, Arizona had a home business mailing out literature for vitamin companies, essentially envelope stuffing. In October 1990, six armed agents and police raided her home, taking all of her business records and literature, putting her out of business. No charges were filed.

Pets Smell-Free made a product to deodorize pets and fish aquariums. The FDA raided the company, claiming the it was selling an "unsafe, unapproved drug". This is their standard claim against anything they don't like. The "vitamin vigilantes" are now also the "doggy deodorant detectives". The FDA seized their entire inventory and all of their business records. The FDA is trying to intimidate them into signing a consent decree which they refuse to do. Traco Labs sold blackcurrant oil. The FDA raided the company and seized their stock claiming that the black currant oil was an "unsafe food *additive*". The FDA lost the case in the appeals court when the judge said that under the FDA's ridiculous definitions, even water added to food would be considered a food additive.

Solid Gold Pet Foods put out a line of holistic pet foods. The FDA didn't like the information that the owner put on the labels. There was certainly nothing wrong with the product. The FDA seized the products from her store *without a search warrant* and shut down her store. She demanded a jury trial. In reprisal, when she showed up for the trial she was shackled with leg irons and thrown in jail for 179 days where she suffered a nearly fatal stroke. She sued the Justice Department and won.

Nature's Way sold evening primrose oil. The FDA seized their stock claiming that it was an unapproved food additive. It is an herb that has been used for thousands of years. Nature's Way sued to get their product back, but were eventually forced to remove *vitamin E* from the oil. The FDA said that Vitamin E was not approved as a food additive to evening primrose oil. It can be added to children's cereal, bread, and everything else, but it had not been approved as a food additive for evening primrose oil which they had claimed was itself a food additive in the first place.

Perhaps the silliest raid of all was against Natural Vision International. They made glasses with no lenses but pinholes in an opaque blank. The purpose of the glasses was to exercise and strengthen the eyes. Opticians and ophthalmologists who didn't want people with stronger vision complained to the FDA until they raided the company, seizing their entire stock and driving them out of business. What kind of food or drug would you call "glasses" with cardboard lenses full of holes? NVI submitted hundreds of testimonial letters from satisfied customers, but the FDA refused to return their merchandise. The medical establishment's punishment agency succeeded again.

The list just goes on and on. The FDA tries to intimidate anyone who publishes any kind of information on healthy products that they produce or sell. This tactic is used to prevent anyone from supplying any information to promote their product, making it very difficult to sell it. When that doesn't work, they raid the company or owners and try to blackmail them into signing consent decrees to avoid the expense of defending themselves. If that doesn't work, they try to get summary judgements out of judges to avoid going to trial, especially a jury trial. This is clear, blatant abuse of police and regulatory powers to protect the industry they are supposed to be regulating.

There are hundreds of other cases of blatant violations of both the intent of Congress and the Constitution. In just one day, the FDA conducted thirty-seven raids of doctors' offices, supplement manufacturers, and health food stores with armed agents. They seized hundreds of perfectly legal products, records, computers, office equipment, and other items. These people had violated no laws and for the most part, were never charged with anything.

The FDA has signed an agreement with the Pharmaceutical Advertising Council to try to destroy alternative medicine, using their police powers to trample the Bill of Rights to dust. Who is the Pharmaceutical Advertising Council? They are representatives of the big drug companies that the FDA is supposed to be regulating. While the FDA has steadfastly denied that they are trying to suppress the use of vitamins and other supplements, their actions say just the opposite. Now they are beginning to admit it. From their own publications comes the following statement. They say that their *job* is "to insure that the existence of dietary supplements on the market does not act as a disincentive for drug development." In other words, if an inexpensive dietary supplement has any prospect of replacing a drug or making the development of one unnecessary, they will take action against the supplement and any information about it. Forget the First Amendment rights of free speech and freedom of the press.

A classic case of the above is the action against L-tryptophan. This product has been used for over twenty years as an effective agent for insomnia and depression. It is a precursor for the manufacturing of serotonin in the body. It enjoyed sales of about $180 million a year because it worked very well. It is an amino acid found in milk and other sources and is completely natural and safe. A Japanese manufacturer had a batch of it that was contaminated. This contamination caused the problems in that batch. The FDA pulled *all* L-tryptophan off the market and has steadfastly refused to let it back on for no reason other than to protect the profits of the drug companies making sleeping and anti-depressant drugs. Doctors can still prescribe the exact compound that was in the health food stores, but many, like my doctor, have been scared off by the negative propaganda from the FDA and the possibility of lawsuits if a problem arises. If you can even get a doctor to prescribe this natural substance, it will cost you seven dollars per pill, versus the ten cents per pill cost before the removal from the health food stores. Are there any questions about the FDA's real purpose?

A good replacement for L-tryptophan is Melatonin. This is a natural substance manufactured by the body to help with sleep. Its production is triggered by the absence of light. It is also very effective for jet lag and now a number of clinical studies show it is also good for a number of other things such as reducing the risk of cancer, particularly breast cancer. The FDA is doing everything in its power to intimidate manufacturers into stopping production, including the threat of raids.

The FDA wants to regulate both single amino acids and mixtures of amino acids, like those you get naturally in meat, as drugs. Millions of people take L-lysine for cold sores and other virus problems. Bodybuilders use amino supplements to facilitate muscle repair and growth. None of these common amino acids would be available if the FDA had its way. You *might* be able to get them by prescription at an exorbitant cost after you pay for a doctor's visit. What they don't regulate as drugs, like herbs, they want to regulate as food additives. Under their definition, even water would be a food additive according to that appeals court judge. Most FDA cases that get to court are thrown out as absurd. But most cases don't get to court. The idea is intimidation, confiscation, and persecution, not prosecution.

The FDA has assumed the right to make their own definitions of things, like Alice-in-Wonderland, and then enforce them. If anyone makes any kind of claim or supplies any information on how a food supplement, including vitamins, can improve your health, that information, even if it came from a government-sponsored clinical study or medical journal, constitutes a "drug claim" and makes the substance a drug. Since it is well-known and often published that eight large glasses of pure water a day can alleviate many health problems (which makes water a drug under their definitions) don't you wonder why the FDA doesn't take action against the many dangerous and unsafe municipal water systems in this country? Believe it or not, they have repeatedly taken action against bottled water companies who even hinted that "city water" might not be safe or that their bottled water was in some way superior. I wonder when air will become a drug.

The FDA does have some legitimate functions. If it would spend a lot more time inspecting food and getting all the illegal pesticides that are still being used out of our food, instead of wasting their resources playing vitamin vigilantes, we would be a lot healthier. Actually, the FDA should be abolished. If it had never been created, a lot more people would be alive today. The food regulation function should be transferred to the Department of Agriculture, where it belonged in the first place. In 1900 there was one Department of Agriculture employee for every 1,800 farms. Today there is one bureaucrat for every sixteen farms. All bureaucracies become wildly growing cancers unless heavily restrained. Let's at least use some of these bureaucrats to make our food supply safer. The FDA is doing a lousy job of that because, in part, they are too busy raiding health food stores and doctors' offices who tell the truth about food supplements. The drug regulation part of the FDA can be provided by a screening group with the following provisions.

Any drug that has been approved for use in, say, three countries, maybe Germany, Sweden, and Japan, (use six countries if you prefer), can be marketed here on a trial basis with required feedback from both patients and doctors for a three year period. If it passes some reasonable criteria, it is approved for general use and the feedback becomes optional. This system would provide far better and cheaper screening and safety than the ridiculous process we have now.

Once a drug company buys the FDA's approval, it is very hard to get off the market no matter how many people it kills or damages. The FDA does not like to lose face by admitting that its absurdly expensive process is as faulty as it is. This process mentioned above would allow smaller drug companies to have a chance to get new drugs approved at a reasonable cost and enter the market. If Hillary really wants to reduce medical costs, all she has to do is tell Bill to eliminate the FDA. Not only would the administrative costs be eliminated, but the price of drugs would drop dramatically. Even more important, alternative and complementary medicine, which are often more effective and a lot cheaper, would expand even faster than they already have, and the total savings would be in the billions. That doesn't even count the lives and money saved through widespread dissemination of the valid information on using nutrition to improve health which the FDA now suppresses in direct violation of the First Amendment.

How does the suppression of information kill people? I'm glad you asked. In the 1950's, the Shulte brothers conclusively demonstrated the ability of vitamin E to help prevent heart attacks and speed the recovery from them at their clinic. Others took up the use of vitamin E, and there are thousands of clinical studies and case histories of its effectiveness. Nevertheless, the FDA and the mainstream medical establishment denounced vitamin E as worthless for this use for over forty years. In the meantime, heart disease became the nation's number one killer. Recently, the FDA has caved in under the mountain of evidence and has grudgingly admitted that Vitamin E *might* be useful in this area. However, one widely quoted "medical expert" has said that the widespread use of vitamin E to help prevent heart disease should not be recommended because "we" don't know the effects of long term usage of doses above the RDA. There is no known toxic level of vitamin E intake. You could probably drink a quart of it a day without ill effects. There are two well known effects of widespread use of Vitamin E that I can suggest to this gentleman. One, fewer people would die from heart attacks. Two, the surgeons and hospitals would make a lot less money from bypass and angioplasty procedures. I wonder if there is any connection between these two effects and the medical establishment's opposition to the use of vitamin E.

Despite all the evidence and their own admissions that vitamin E can be useful, if the FDA can somehow entice a health-food clerk into saying that vitamin E might help prevent heart attacks, they can confiscate everything in the store and shut it down. See the Survival Manual chapter for more details about the kind of information the FDA is trying to suppress.

The FDA/AMA/drug cartel is trying desperately to shut down complementary and alternative medicine, despite the fact that the treatments are often more effective, safer, and much cheaper. What we now call conventional medicine is only about fifty years old, dating from the advent of antibiotics. These people feel very threatened by alternative and complementary medicine. If these approaches did not offer real competition because they worked, the medical establishment would not feel so threatened.

INTERNAL REVENUE SERVICE

It would hardly be news to anyone that the IRS keeps getting out of control. The stories of their abuses would fill several books. Congress had to pass a Taxpayer's Bill of Rights to try to stop some of the abuses, but that law has not succeeded. The IRS is under enormous pressure to increase revenues; as a result, the abuses will only grow. To help increase revenues, Clinton wants to increase penalties by twenty-five percent, hire 21,000 new auditors and increase the number of audits to five times as many as there are now. So far, Congress has authorized 5,000 new enforcement agents.

When you are called in for an audit, the IRS agent is expected to get more money out of you so as not to waste his time. Sometimes they go to ridiculous extremes to do so. They were going to seize a company called Rainmaker, for a *one-cent* underpayment of the firm's withholding taxes. A Congressman had to intervene to stop the seizure. There are those confiscation laws again.

A sixty-five-year-old doctor decided to consolidate his various bank accounts for convenience. The IRS immediately confiscated every penny, because they claimed he had failed to fill out a form required by them for moving cash or deposits around. He didn't know about any form. The bank apparently didn't know about the form. All of that didn't matter; the IRS took every penny. The District Attorney didn't even leave him enough to hire an attorney, which is a common trick in confiscation cases. He can prove he earned the money. He can prove he paid taxes on it. He has every right to the money. It was perfectly legal for him to have and keep it. If he had moved it in smaller amounts over a period of time, there might have been no problem, except that there is also a regulation about using multiple transactions to avoid the reporting requirements. None of that matters. His money which he worked for all of his life is gone. He didn't fill out a form that virtually nobody knows anything about; now he can't afford to retire and will probably never be able to. Everything he worked for all through his life, is gone. Such incidents are only the tip of the tip of the iceberg that will eventually sink many of us when the economy collapses and the government gets even more desperate for money.

There is a saying that ignorance of the law is no excuse. That may have been true a century ago when most laws were based on the Ten Commandments which virtually everybody was familiar with. You had to deliberately try to violate a law. Today, there are over 40,000 regulations dealing with McDonalds serving you a Big Mac. It is virtually impossible for them to do so without breaking some government regulation somewhere in the process. Now it is almost impossible for you to get up in the morning, go to work, come home, and go to bed without violating some law or regulation. It's no wonder that people have no respect for laws anymore. You can't help but break a bunch of them almost every time you do anything, so why worry about breaking any of them?

The IRS has come up with a national identity card. It will contain all of your employment, income and spending information so they can make sure everything matches. Welcome to Orwell's 1984. He was a little off in his time scale, but not in the eventual results. The IRS has got to go, at least as we know it now. The only way to do this is to get rid of the income tax.

The IRS maintains a master file on every taxpayer. It is usually updated every year. If you would like to know everything they know about you, just file a Freedom of Information Act request and they have to send you a copy of your file. Supposedly, depending on your timing, you can get up to a six-month warning if you are going to be audited. If nothing else, 50 million taxpayers filing such requests each year would chew up their resources significantly and maybe slow down some of their harassment of honest taxpayers. You can obtain information on how to file this request with the forms needed from the Life Extension Foundation listed in the Appendix. Filing these requests can be an effective protest weapon against any government agency if enough people join together to do so.

POST OFFICE

Like the IRS, the stories of post office foulups and abuses would fill several books. Almost everyone has had problems with them somewhere along the line. Every few months there is another scandal, discovery, or disgruntled employee shooting up the place.

One such story involves the post office going around fining people for using overnight delivery from competitors. They claimed that overnight delivery was not needed in some cases so the letters should have been sent through the regular mail. First of all, how did they know what was sent? Next, what business was it of theirs? Then, what criteria did they use to decide whether something needed to be sent overnight or not? Isn't that a judgement call by the people who have to pay thirty times as much money to use overnight mail? Don't they know a lot more about their business and the needs of their business than the post office does? What if they just wanted to make sure it got there? Don't they have that right, especially given the post office's reputation? Lastly, you can bet the post office would not have said a word if the material had been sent by overnight mail through the post office. It was an attack against their competitors by an agency that abuses its governmental powers to stifle competition.

This is a blatant attempt at intimidation by an outfit that can't compete with private enterprise, despite having billions of taxpayer dollars in buildings, trucks, and equipment provided to them for nothing. It is also a clear abuse of power. It is time to open up the entire spectrum of postal service to private enterprise. If the postal service can't compete despite their enormous capitalization advantage, then let them go under. Their service is obviously too expensive and inefficient. Such competition is certainly in the public interest, as competition always lowers cost and improves service. It would also be a fitting "penalty" for such flagrant abuses of power.

Since protests are a protected right in America, unless you are protesting abortion, the recent raise in postal rates was a golden opportunity to protest post office services and costs. Remember all those lines of people at post offices across the country trying to buy three cent stamps to upgrade their twenty-nine-cent stamps? It seems that the post office badly underestimated how many such stamps would be needed. What do you think would have happened if the first people who got there bought all the three-cent stamps each post office had? There would have been hundreds of thousands of angry letters, and calls to Congress with many of them demanding that Congress open mail delivery up to private enterprise.

What on earth would someone do with all the three-cent stamps he could buy? Well, you put eleven of them, or ten with one two-cent stamp if you want to bother, on every letter you mail. That forces the post office to sort out and hand cancel every one of those stamps. If a few million people had done that, it would have really snarled things up and raised even more demand for some competition. You might remember this idea the next time the post office decides to raise rates again.

BUREAU OF INDIAN AFFAIRS

> *"Powerful nations and empires have always been alike in one way: they brush the records of their mistakes and*

misdeeds under the rug, where they remain until the searchlight of history uncovers them. Even then, they are never more than footnotes in the records for the country where they occurred." - Chief Red Fox of the Sioux.

The U.S. is, without doubt, the most generous nation in history. No other nation has ever conquered other countries and then rebuilt them and set them free to compete against their conquerors. No other country has given away trillions of dollars of their taxpayers' money to help other countries – well, actually, to help the CFR/TLC agenda. However, there are two great blots on America's history. One is slavery/discrimination, primarily against blacks, and the other is the genocide against the Native Americans of the last century followed by the inept and often corrupt Bureau of Indian Affairs.

The Indians were killed in battles defending their homes and decimated by diseases brought from Europe. Whole villages were massacred. Entire tribes vanished. Many treaties were signed granting them various areas of land that they owned in the first place, and then broken when someone else wanted the land. Food, medicine, and other promised supplies never arrived and were often sold to someone else by corrupt agents. Many of the Indians died of starvation or lack of medical care.

The buffalo that the plains Indians depended upon for food and clothing among other things, were slaughtered to drive the Indians onto reservations and make them totally dependent on the government for survival. If you want a good idea of what the nanny (socialist) government we are headed toward can produce, look at the conditions of the Native Americans today after a century of government care.

They have the highest abortion rate of any group. When Margaret Sanger founded Planned Parenthood, her main stated goal was to limit the birthrate of the "unfit" (her words) like Indians and Blacks. Today these two groups have disproportionally high rates of abortions, as intended, with the government's help.

Given the almost hopeless situation many Native Americans find themselves in, it is no wonder women don't want to bring babies into the world. Alcoholism is a serious problem as it always is when people are stifled, limited, and have no incentive and little hope. Historically, the education provided has been dismal at best.

On the other hand, Bill Clinton has pledged $600 million of your tax money for economic development for Communist South Africa. Hey, Bill, how about some money for economic development on the reservations?

Any group that is "kept," whether in the Chicago ghettos or the Indian reservations, develops a lot of problems. The economic disincentives to any change in status or any hope of improvement tend to trap people into a life of hopelessness, alcohol, drugs and crime. Fortunately, tribal and family influences have helped keep the crime rate down among the Indians, but the other results are widespread.

Most of the tribes and reservations are actually "foreign" sovereign countries by treaty. Some even issue their own passports. They don't have to secede from the Union, as they already are their own countries by treaty and historical right. Many of the tribes have been given "captive nation" status by the U.N., which, in effect, they are. I strongly disagree with almost everything the U.N. does, but they got this one right.

In my opinion, the best thing most of these groups could do is become truly independent countries. In return for what has been taken from them by force, the government should pay reparations for a limited time to be used only for economic development. If we can rebuild and set free Japan and Germany, who attacked the U.S., surely we can do the same thing for the people we drove off their own land. We can't undo the massacre at Wounded Knee any more than we can undo the massacres at Ruby Ridge and Waco, but we can help the Native Americans recover, to some extent, from what the government has done over the last two hundred years.

For the plains Indians, raising buffalo would be a way of providing meaningful employment and a return to much of their traditional life. These new "countries" could also provide international banks with real secrecy that the Swiss banks have largely given up under pressure from the IRS. Larger reservations could have international airports with duty free stores. Some could attract manufacturing companies with various incentives.

Except for raising buffalo, most of the economical development options would tend to take them away from their ancient traditions and way of life, but that is a decision for the Native Americans to make. One thing is sure: the BIA has no business being involved in what the Native Americans do with/on their own land, as is the case now. As long as their dependency on the government continues, the Native Americans never will be truly free or prosperous.

CONFISCATION

> *"The moment the idea is admitted into society that property is not as sacred as the laws of God, and that there is not a force of law and public justice to protect it, anarchy and tyranny commence. Property must be secure or liberty cannot exist."* - John Adams

Unfortunately, your property is no longer secure. Some government agency can take it any time they want to and there is nothing you can do about it. For all practical purposes, the Fifth Amendment has been blown away. They may have to trick you into letting them in your house, or just break in on some "anonymous" tip they supposedly got so they can search your home until they find something wrong. Heaven help you if you have a gun in your hand when they break in without announcing themselves and you try to defend yourself against these unknown, masked, home invaders. You are a *dead* duck.

When all these unconstitutional confiscation laws started being passed, they were to permit seizure of boats, cars, airplanes, and warehouses involved in drug smuggling. Today, thousand of innocent families have lost everything they have without even being charged with a crime, much less convicted of one. Often they can't even find out why they were raided, but *everything they had is gone.*

A family in New Jersey was awakened to find government agents with a moving van surrounding their house. The wife had been accused by a neighbor of stealing a UPS package. There was no real evidence and the wife had not been charged with anything. It was the dead of winter and they were thrown out into the street with nothing but the clothes on their backs. The confiscation crew took everything they had including the house, cars, and all personal belongings. Apparently, the government has never heard of due process.

Then there is the "seize their homes" crusade by the government. If you have or had an FHA-or-VA guaranteed loan, you could lose your home, even if you have paid it off. How? By having made the slightest error in filling out the original form.

As Daniel Rosenthal reported, "Thousands of other families are losing their homes because they allegedly provided false or inaccurate information – purposely or not – in applying for their mortgages. It doesn't matter that you've paid your mortgages. It doesn't matter that you've paid your mortgage on time every month. All that matters is that the regulators come to believe the form is wrong. If your banker believes your income was overstated on your application, he is required by law to report it."

In other words, if you were earning some amount like $27,400 but thought you were going to get a raise before the year was out and rounded the number off to $28,000 on the application, but your income tax return for that year only showed 27,700 because the small raise you got was near the end of that year, your house can be (and people's have been) seized and forfeited for overstating your income. If you were on a commission or bonus plan and had to estimate your income, Heaven help you. The government is currently in the process of cross-checking every home any agency had anything to do with. It will take years, but they will get to yours. You had better hope that they can't find anything wrong.

These laws have to be changed or eliminated. At least there need to be some iron fences built around them. Property should only be eligible for confiscation *after* a conviction and the appeals have run out. That's called due process (Fifth Amendment), which our lawmakers and courts seemed to have forgotten about. Giving the police, prosecutors, and judges a monetary reward for taking your property is *guaranteed* to generate massive abuses. This is nonsense. The system is corrupt enough without this kind of incentive to trample on people's rights.

If you think you are safe from such antics, think again. All someone has to see is a wild marijuana plant growing on your property that you know nothing about, report you for a share of the booty, and your house and property are gone. There is no recourse. If some neighbor gets mad at you he can even plant the marijuana and then call. Poof! Everything you own is gone for good. If he just wants to hassle you, there's the anonymous tip that will have some agency breaking down your door, usually at dawn. If they find something, you're history. If not, they usually will have torn your house to shreds looking for something, as in the previous stories. If some agency wants to get you for whatever reason, they just plant the evidence themselves or bring it with them to "find" on your property as was the situation in numerous Boston cases. This is ludicrous. We have to put a stop to it or there is no such thing as private property rights anymore. As John Adams said above, property must be secure or there is no freedom.

If you think such things are far fetched, Gerry Spence, one of the most respected defense attorneys in the country, told the Montana Trial Lawyers Association that he had never been involved with a case with the federal government in which the government had not lied and manufactured evidence to gain a conviction. Recently, even more information about the FBI manufacturing evidence has been reported in the news.

To give you an idea of a common bureaucratic mindset, one high ranking bureaucrat recently stated that "ownership of private property is a quaint anachronism." In other words, forget the Constitution.

Let me close this section on bureaucracies with a quote from that lawyer, Gerry Spence, who successfully defended Randy Weaver by proving that the government was lying, manufacturing evidence, suppressing evidence, and knowingly charging the wrong person. This was confirmed by the Justice department's own investigation. In case you miss the point in his quote below, the wolf is our own government. (The words in parentheses are mine, not his.)

> *"Although we give lip service to the notion of freedom, we know the government is no longer the servant of the people but, at last, has become the people's master. We have stood by like timid sheep while the wolf killed – first the weak* (Weaver's?), *then the strays* (Waco?), *then those on the outer edges of the flock* (those who speak out against the government – Omnibus Anti-terrorism Bill), *until at last the entire flock belongs to the wolf."* - Gerry Spence

This sounds hauntingly like the quote about what happened in Nazi Germany in the first chapter.

EQUAL EMPLOYMENT OPPORTUNITY COMMISSION

The EEOC has a legitimate but limited role to play in seeing that people's fundamental rights are protected. As usual, the bureaucrats have run off of the cliff with their power. The latest attempt to impose political correctness (humanism in this case) was the proposed regulation concerning freedom of religious expression at work. It stated that people could not have Bibles on their own desks; neither could they wear any jewelry to work that had a religious theme, such as a cross, not even on a charm bracelet. This is a total trampling of the First Amendment, but such things as the Constitution are irrelevant to many of our bureaucrats. Thanks largely to Christian radio, which the liberals are trying to shut down, the EEOC received more than 100,000 letters protesting the proposed guidelines, and has backed down for now.

Some of the decisions of the EEOC have been absurd. Businesses have shut down because they couldn't possibly comply with mandates resulting from complaints that were obviously phony to start with. It's almost like some agencies, or at least some of the administrators, want to collect as many scalps as possible to prove they are doing their job. The disabilities act has given them a whole new arsenal to attack businesses. While there is a legitimate need to protect disabled people's rights, the definition of someone with a disability in the act is so broad that virtually anybody can think up a disability. One man filed a complaint when he was fired for threatening fellow employees with a gun. He filed a complaint under the disabilities act claiming that he was mentally disturbed and therefore disabled, and they couldn't fire him. Don't laugh; he may win.

As usual, Congress passed a law with sweeping generalities and the bureaucrats turned a molehill into a mountain of regulations and red tape. How can we compete in a global economy when businesses are buried under tons of regulations and requirements, and thus have to spend most of their time looking over their shoulders or trying to find out what the regulations are, much less complying with them? History has proven that some regulation of commerce is necessary. That level of necessary regulation is about one percent of what we now have. Each year the bureaucrats crank out enough regulations to fill 68,000 pages of the Federal Register in very fine print. I wonder if the world would come to an end if we skipped a whole year's quota of these regulations. This deluge has got to stop.

CONCLUSION

The massive bureaucracies that have grown up to implement the nanny government have become laws unto themselves. "The Mission" is all that counts. People's rights only get in the way of the mission. The only way to clean up the mess is to restrict the federal government to only that authority which the Constitution allows. The only way to do that is to elect a whole new government consisting of people who will honor their oath of office to uphold and defend the Constitution. The ones we have now have proven they won't do it.

LATE INFORMATION

I sincerely hope that the following information does not belong in this chapter. However, some very disturbing questions have come up as this book goes to print about the Oklahoma City bombing that need to be answered.

According to at least a dozen military demolition experts, two bombs had to go off almost at the same time to cause the observed damage to the building. This has been confirmed by seismograph recordings at two different locations, one in Norman Oklahoma and one in the Kirkpatrick Center. These recordings clearly show two nearly equal explosions at the building about ten seconds apart. These cannot be echoes, reflections, or anything else but two separate explosions. These recordings were briefly mentioned by the media shortly after the bombing, but the information was suppressed by the government. Why? Whatever happened to independent investigative reporting?

According to all of the explosives experts, the truck bomb could not possibly have caused most of the damage to the building. It was much too far away. A larger bomb of the same type was exploded *inside* the World Trade Center but caused far less damage, even though most of the force from such a bomb goes straight up, not to the sides. The building was built with reinforced support columns to withstand such an incident. At the World Trade Center, the bomb went off right beside such a column and did not take it out. The Oklahoma bomb was much too far away to even slightly damage most of the columns, but many of them were completely cut in half. According to the experts, that could only have been done by a series of special devices wired to *each* of the columns.

The damage to the building was in the wrong places. The damage should have been in a cylindrical pattern around the bomb. It was not. Columns that should have collapsed if the explosion had been strong enough, did not, while much heavier columns much further from the explosion were neatly cut in half. This has to have been an inside job, since this was a high security building. Nobody could have just wandered in and planted the special linear charges that had to have been used to cut the columns like they did. There had to be two teams, one with the truck and one placing the explosives inside the building.

In addition, people listening to police scanners heard about a third bomb found in the building unexploded. Rescue efforts were suspended until a military bomb squad could disarm it. The bomb reportedly had military markings on it and was a sophisticated fulminated mercury bomb. Had it gone off with the others, the whole building would have come down. This bomb was also briefly mentioned in news reports shortly after the explosion and then suppressed by the government. They deny there was any such bomb. Why?

Soon after the explosion, the American news media began speculating about a Middle Eastern connection since the bomb was virtually identical to the one at the World Trade Center. *Immediately*, Tehran radio and TV stations announced that it was the work of right-wing Christian militias. *How did they know* that it was Americans when we had no clue yet unless they had inside information? Was McVeigh an unwitting decoy? The two bomb teams did not have to know about each other. If they didn't, who coordinated the explosions? They missed in their timing by about ten seconds if the explosions were supposed to go off at the same time. The truck was parked in the wrong place if it was to be blamed for all the damage.

Even nastier questions keep surfacing. At 9 a.m. on a Monday morning when the BATF offices are normally full, why was the place virtually empty? According to later reports, the records about Waco were stored in these offices and were expected to be subpoenaed for the congressional hearings. When the BATF offices came through relatively unscathed, the government *stopped* the rescue efforts for several hours while BATF agents carried these records out of the building. *Why? Where are the records?*

One of the most puzzling actions by the government was the extreme hurry they were in to demolish the building and destroy the evidence, just like they did at Waco by burning the building down and bulldozing everything into the basement, and covering it up. McVeigh's defense attorney tried to delay the demolition so he could get some independent experts in there to examine the evidence. The government would have no part of that. They were in such a hurry that they set a deadline on the rescue efforts, so that they could destroy it before anyone could examine it. There were still two bodies missing as the deadline approached. Rescuers wanted to postpone the demolition until they found the bodies. No way. Fortunately, they did find them just before the deadline.

A careful examination of the damage would have given an incredible amount of unique information on how the damage occurred, the damage pattern of the bomb(s), how to build more bombproof buildings, and a whole lot more. That damaged building was invaluable. Why did they have to demolish it before it could be properly examined? The excuse about public safety just doesn't hold up. It wasn't about to fall down. The uncut supporting columns were far too strong for that. Besides, nobody was allowed anywhere near it except for rescue workers and certain government officials. Somebody had to be in a desperate hurry to get rid of the evidence. *Why?*

According to the bomb experts, the truck bomb had to be set off manually, on the spot, by someone in the truck. No one saw anyone approach the truck and open the tailgate before the explosion. The security cameras did not record anyone approaching the truck. Later, it came out that there were excess body parts near the truck that did not belong to any of the known victims. Apparently, the blast was set off by a suicide bomber who remained in the truck for some time after it drove up. Americans usually don't go in for suicide bombings like the fundamentalist-religious fanatics from the Middle East do.

There are many possible scenarios for the explosion, but one of the most likely is as follows. The explosives to cut the columns were set off first so that the building would be the most vulnerable to the other blasts. The mercury fulminate bomb was to go off next, taking out the rest of the supports, while the truck bomb went off last, to blow the debris back into the building and look like it caused the damage to cover up the inside work. That would explain the ten-second delay between the blasts. Even if there wasn't a mercury bomb, as the government now claims, the scenario worked as intended, as the FBI continues to insist that only one bomb went off. Why would anyone want to cover up the fact that there were two different explosions?

Was there a Middle East connection? Was there a U.S. government connection? Given the massive government cover-ups in the past, we must demand some straight answers. These are not wild speculations mentioned above, but facts and expert analysis. Seismographs do not lie; government agents do.

One of the nastiest questions is the obvious one to ask. Who benefited the most? Clinton's dismal and falling rating shot up over ten points as he shamelessly used the tragedy for his own political posturing. The anti-terrorism bill that gives him and various law enforcement agencies, like the FBI and the BATF, broad, sweeping new powers in violation of the Constitution was dead in Congress until the bomb went off. With this bill, he can designate anyone or any group as a terrorist for any reason. He can suspend their Constitutional rights and jail them without charges or any form of due process indefinitely. His actions are *not* subject to review or interference by *any court.* This is unheard of in America. Why are these provisions in this bill? He can classify any group that opposes his policies, for example, Operation Rescue, as a terrorist group. If you donate any money to a local pro-life group who sends some of that money to Operation Rescue without your knowledge, you could be jailed for many years without a trial. If you can't believe that, read the act. Was McVeigh a stooge like Marimus van der Lubbe in the Reichstag fire, which was actually set by Goering, that allowed Hitler to pass similar legislation that he *first* used against the Communists? Everything in me cries out against such a thought, but with what is at stake, we must demand that ALL the facts be brought out and all the questions answered.

(A significant amount of the information in this chapter was found in various issues of *Soldier of Fortune* Magazine, *Life Extension* Magazine and Internet newsgroups.)

Chapter Three
It's the Stupid Economy

"Permit me to issue and control the money of a nation, and I care not who makes its laws." - Mayer Amschel Rothschild

"All the perplexities, confusion and distress in America arise, not from defects in their Constitution or Confederation, not from want of honor, or virtue, so much as from downright ignorance of the nature of coin, credit and circulation." - John Adams

Probably not one person in ten actually understands what real money is. Far fewer understand how our fractional reserve banking system works. If you are going to protect yourself, much less profit from the coming economic collapse, you need to understand both.

For thousands of years, money was something that had a known intrinsic value. Usually it was made out of silver or gold. The material was valued for uses other than money. It had to be found, mined and refined. It took work to own it. Thus, money had a certain intrinsic value of its own apart from what the government or whoever might stamp on it.

Carrying or safeguarding large amounts of money in this form was not very convenient, so depositories sprang up where you could store your gold and receive a receipt for it. Any time you wanted some of your money, you took the receipt to the bank to redeem it. Originally banks charged a storage fee to pay their expenses. Then they figured out that they could loan out some of the money on deposit and charge interest, making even more money. If the loan was repaid, all was well and good. If not, the bankers had to replace the lost money. Bankers were very careful about whom they loaned money to in those days when there were no government guarantees.

While carrying around a paper receipt was more convenient than the coins, it was still a bother to go get the coins whenever you wanted to make a big purchase. So people just started trading the receipts instead. Soon very few people ever came in to get the gold or silver coins. At this point, the bankers figured out that they could loan out more money than they had on deposit. Since so little of the real money actually circulated, nobody would know the difference. Thus fractional reserve banking and the inflation that goes with it was born. The bankers were creating "money" out of thin air by loaning it into existence by way of phony receipts. They still are, only today the federal reserve banks create trillions of dollars out of thin air to finance the governments deficits. That is why the dollar has lost ninety percent of its value since the Federal Reserve Act of 1913.

Between 1800 and 1913, the value of a dollar bounced around a little but basically was nearly the same over the entire period. Then Congress abandoned their responsibility and accountability for the nation's money by passing the Federal Reserve Act. This created a number of privately owned "central" banks to control the nation's money supply. The consumer price index (inflation) started rising immediately. That rate of growth was relatively slow until President Johnson was forced to violate the government's promise to redeem its money, specifically silver certificates, in real money made from silver. The deficits he brought about had caused so much phony money to be created that people started turning in the intrinsically worthless paper for real money, and there wasn't enough silver.

> *"The entire Federal Reserve System and the fiat currency on which it is based is a scheme founded on deception and dishonesty on a scale so massive that its premises are seldom challenged."* - Howard Phillips, National Taxpayers' Party

The Consumer Price Index and inflation really took off after Nixon had to stop exchanging gold for dollars with the banks of other countries we had flooded with our paper promises. At this point, there was no longer any connection between what was printed on the paper and the real world. In 1920, a $20 gold coin would buy a very nice suit and a whole lot of other things. Today, that same gold coin will still buy a nice suit and those same things, but now that same coin is "worth" about 400 paper dollars.

That inflation (loss of purchasing power) in those paper dollars is due to our government's deficit spending. Unwilling to pass the necessary taxes (as bad as they are) to pay for all of the politicians' schemes and pork projects to keep themselves in office, because the public would never stand for it, they simply borrowed the money, which the banks created out of thin air. In simple fashion, here is how it works: the treasury prints up some treasury bonds. Cost? Zilch. They give these pieces of paper to the federal reserve banks and others as "promises" to pay them back, which they never actually do. They just print more bonds and borrow more money. They have, so far, redeemed them when the specific bond came due by effectively rolling them over into new bonds that they can't redeem any other way either. So, the total debt just keeps on growing.

In return for the bonds and the interest, the Federal Reserve Banks then deposit the "money" that they just created out of nothing into the government's accounts. At this point it gets even worse. Now the banks add those bonds to their "reserves" as assets and loan far more money to people based on whatever the reserve requirements are at the time, creating more money out of thin air. Thus, when the government borrows money into existence, rather than paying it into existence (still phony money either way), the inflationary effect is even worse.

Abraham Lincoln ran into a similar situation during the Civil War and bypassed the bankers by issuing *non-interest bearing* United States Notes that he "paid" into existence. If you are old enough, you might remember some of these notes with red seals on them. Shortly after he announced that he was going to continue this practice after the war, he was assassinated. In a speech in 1963, John Kennedy announced that he was going to return to Lincoln's practice and bypass the banks as the Constitution originally required. Shortly afterward, he too was assassinated. Sort of makes you wonder, doesn't it?

Is this another wild conspiracy theory? Perhaps, but if you think that such an idea is utterly absurd, the Bank of England was extremely upset with Lincoln's plan to issue money directly from the Treasury as the Constitution intended. Here is what the London Times said about his plan:

> *"If that mischievous financial policy, which had its origin in the North American Republic during the late war in that country, should be come indurated down to a fixture, then that government will furnish its own money WITHOUT COST. It will pay off its debts and be without a debt. It will have all the money necessary to carry on its commerce. It will become prosperous beyond precedent in the history of the world."*

You might want to go back and read that again, slowly. This was the forecast from people who were bitterly opposed to Lincoln's plan, not some starry-eye supporters. It's no wonder Lincoln had to go.

If Lincoln's practice had been continued instead of creating the Federal Reserve System and money had been paid into existence (created) by the government at exactly the same rate as the increase in productivity since then, the value of the dollar would have remained stable and we would have no national debt. We would need no income taxes, and we would be the wealthiest people on earth, because of our tremendous gains in productivity. Instead, the politicians abandoned their responsibility and turned it over to private banks. They then spent everything they could borrow and tax on their various social schemes. The bankers made ridiculously bad loans all over the world to try to gain control of everything they could. As a result, you and each member of your family "owe" about $17,000 to someone, mostly large banks, when you didn't even borrow the money or receive any benefit from it. Are you ready to make a change? The Federal Reserve System has to go.

One of the stupidest policies ever carried out by our government was to deliberately drive down the value of the dollar in international markets when the dollar was the king of currencies. If the dollar was in fact overvalued, the natural corrective forces of a free market would have made any reasonable corrections. Instead, the liberals/socialists who think they can control things much better than natural free market economic principles can, intervened. At the time, we were the greatest creditor nation in history. Virtually everyone owed us money. The excuse for driving down the dollar was to make our products more affordable to other countries so that we could export more and discourage imports in order to improve our balance of trade deficits. What utter nonsense. Behind the scene, the one-world socialists knew full well that they had to bring America to its knees economically to the point that we were willing (forced?) to join in the one-world government.

Let's look at the results of this policy. At the time, we were the largest creditor nation in the world. Our currency was preferred over all of the others around the world. Interest from the money we had loaned to other nations was flowing back into this country. We had the highest standard of living in the world. Since then we have become the largest debtor nation in the world. At that time the Japanese yen was about 300 to the dollar. Today less than 100 will buy a dollar. That means the dollar you earn today is worth one-third what it was a few years ago, despite the precipitous drop in the Japanese stock market and other problems they have had recently. Have these policies solved our balance of payments problem? Of course not; it has only caused the costs (in cheaper dollars) of all those imports to soar.

The rising prices of oil and other items has only made the problem worse. When OPEC switches to requiring payment in other currencies, the dollar will cease to be the world's reserve currency. Being the reserve currency is all that is supporting the dollar at the moment. When that happens, the value of the dollar will fall like a rock. You had better be prepared for the consequences or you will probably lose most of what you now have, including your job, as our economy collapses. It will happen. Even worse, Clinton and his cohorts are going to come out with a two-tiered dollar, one color for domestic consumption and another for foreigners. Their excuse is that this scheme is to foil counterfeiters and money laundering. Balderdash! It is to try to keep the economy going through the next election and force people to use fully traceable money for their transactions. Throughout history, any such two-tiered system has been a frank admission of desperate economic problems and the value of that country's currency has flopped. That is what is about to happen to us.

The only chance we have to maintain any kind of stable money supply is to return to a gold standard. Governments hate gold because it reveals their chicanery. Central banks hate gold standards for money because it prevents them from creating it out of thin air. Yet they keep their vaults full of gold, as they know full well that it is the only *real* money and that its price cannot be easily manipulated.

The net result of creating an inconceivable amount of phony (fiat) "money" out of thin air has been a substantial lowering of the American standard of living. What the government actually was saying to us was that they were going to deliberately make the money we were earning and had saved worth less, so that they could fund their socialist social policies and thereby buy the votes to keep themselves in power.

Think about this. If they make our money worth less, don't we have to work harder to make more of the cheaper money to buy the same things? That is exactly what they did and continue to do. Today the standard of living of most people in America is less than it was twenty years ago. Yet our productivity has increased considerably. Who got all of the benefits from our hard work? Many if not most families, particularly younger ones, have to have two wage earners to survive, and our children are bearing the brunt of this profound change in our families. More people live below the poverty line than in the past, despite the liberals/socialists spending trillions of dollars on ineffective social programs to buy poor people's votes to keep themselves in power.

Now, fully eighty percent of the jobs in this country are in service industries. That means that only twenty percent of the working people in this country are actually producing anything, which is what maintains a country's wealth. What this means is that we are mostly working to serve one another and buying our products from other countries. Unless this is reversed, along with the continuing fall of the dollar, we *must* soon suffer an economic collapse and essentially become just another third-world country. There is nothing that guarantees America's continued dominance in world affairs, particularly the dominance in economics which it has already lost. We used to have the largest banks in the world. Now we don't even have one in the top ten.

The only thing that has kept our situation from deteriorating much further is that the dollar is the world's reserve currency. Gold, oil, and other commodities are still priced in dollars. Also, foreigners hold an enormous amount of money in dollars and have a vested interest in maintaining its value. That will soon come to an end. The one-worlders are planning on replacing the dollar as the world's reserve currency by the year 2000. When that happens, as it must with our present policies, our standard of living will drop like a rock. Our currency will be just another piece of worthless paper backed by nothing except the government's promise to redeem it with more worthless paper.

When Lyndon Johnson decided to remove the silver from our coins, he broke the last internal link between our currency and real money. The fact that the value of the dollar had dropped to the point where the silver in the coin was worth more than the value the coin represented should have been a screaming warning to change our policies and institute financial and government reforms. Instead, he took the easy way out and debased our currency to continue funding his social fantasies and buy votes, just as the Romans did many centuries ago. Remember guns and butter? The end result of this kind of policy, which still continues, must be the same as it was for the Romans – economic collapse.

Let's take a look at the example of Japanese policies. They welcomed America's foolish program of reducing the value of the dollar and increasing the value of the yen. Japan is a very resource poor country. They have to buy most of their raw materials from other countries. They emphasize products with high value and low raw material cost, such as consumer electronics. They also emphasize added value to the raw materials they do import, such as quality in cars. With a high value yen, they can import the raw materials they do use more cheaply. Doesn't the high-value of the yen make their exports harder to sell in other countries? Yes, but there are remedies for that. For one thing, the higher-valued yen makes it much cheaper for them to buy real estate and companies in other countries. They can buy or build the manufacturing facilities in countries with very low labor and other costs. The profits come back to Japan, raising the standard of living of the whole country. They can loan their overvalued money to other countries as we once did, with the interest also flowing back into their country.

Our liberal/socialist government has been and continues to deliberately commit financial suicide which must inevitably lead to the loss of our freedom as well, which is its ultimate goal. The privately owned Federal Reserve Banks are now trying to defend the dollar from the results of their previous massive creation of dollars out of thin air, in order to try to fund our government's out-of-control spending by raising interest rates. Who is paying the price for this policy? We are, of course, in the form of higher interest rates and a slowdown in the growth of our economy. The current drive for free trade agreements is just one more step in nailing the lid on our financial coffin.

The Trilateral Commission (along with the CFR and others) is behind the current drive for "free" trade agreements, supported by the large multinational corporations and banks. If anyone thinks that these policies are going to increase the standard of living in America and result in more jobs, s/he has his or her head in the sand. The primary purpose of these trade agreements is to reduce the sovereignty of the countries involved, particularly the U.S., with the goal of eventually establishing a one-world economy ruled by the socialist/intellectual elite. The secondary purpose is to facilitate the movement of jobs by the multinational corporations from high cost areas (U.S.) to low-cost areas and still be able to import those products back into the US without paying any duty. The American production worker is being shot in the head and the rest of us will find out that we have been shot in the stomach.

> *"There is nothing like a crisis to bring out the true character and mettle of a people. When the pressure is on, people either pull together or they tear each other apart. A people's spiritual values, the laws which stem from these religious roots and the individual and collective behavior patterns taught and applied, determine largely whether a people will pull through together during rough times, or self-destruct."* - R. E. McMaster, Jr.

During the Great Depression of the 1930s, people pulled together and helped each other. We were still a rurally-based economy. Today, we are largely a dependent urban culture whose moral values have been under vicious attack, and we are changing to a survival-of-the fittest (best armed and organized) and everyone-for-themselves mentality. During the upcoming economic collapse, the change in foundational values will be evident as some of our cities go up in flames.

DEFINITIONS:
Capitalism: savings from past and present applied to the future.
Debt Capitalism: borrowing from the future to spend in the present.
Socialism/communism. destroying the future (and freedom) to build a "better", more just (as in the Soviet Union) society.

We have one foot in the camp of debt capitalism and one in the camp of socialism. Is there any wonder that we are losing control of our choices?

Democrats have been in effective control since 1932. They gave people what they thought people wanted. They promised to protect them from serious want, provide security against their own folly, eliminate risk, and establish universal welfare and medical care. People believed them and sold their freedom to be taken care of, voting them in again and again. Some now realize the folly of believing the government from the pain they suffered, the lack of freedom, government restrictions, and being trapped in poverty. Government rules keep people in economic dependence (slavery?) by punishing incentive and effort. The politicians believe that if you provide what the people want, the people will give the politicians what they want – power. Unfortunately, socialism (big government) cannot deliver on its promise, because it is based on false premises and a lack of understanding of human nature and basic economics. It is hopelessly inefficient and destroys incentive.

What has it cost us for the politicians to buy their continuance in office with our own money? It takes $1 billion a day to pay the interest on the national debt and the deficit grows by another $1 billion each day. In other words, we are borrowing the money to pay the interest. This is a sure road to disaster. As the interest continues to grow, so does the borrowing to pay it. Soon we will be in a runaway situation. The U.S. has $19 trillion in total assets, including everything you own down to your socks. The public and private debt is $20 trillion. In other words, the United States of America is bankrupt and can't even pay the interest on its debts without borrowing it.

While one report from the Bureau of Labor Statistics claimed that 465,000 new jobs had been created in March, a second report or theirs showed that there were 221,000 fewer jobs in March than there were the previous month. Either the Bureau is grossly incompetent or part of it is deliberately lying for political purposes, probably both. Since 1992, the number of government workers in America has exceeded the number of manufacturing workers. Is it any wonder that the majority of Americans think that the government is out of control? Is it any wonder that all these government workers are regulating our freedom out of existence?

Columnist Jack Anderson expects hyperinflation to strike America:

> *"South American sources tell me that our addiction to debt spending is leading us down the same tragically flawed path they traveled, a path that leads to hyperinflation.... Though every fiber of my being wants to believe that we can maintain monetary stability, and sustain an economic recovery, I am compelled to agree with my sources' conclusions. The danger is real and immediate... America is in imminent danger of South American, Banana Republic style hyperinflation... if inflation does return in earnest to the US, the experience of South Americans shows it can strike fast, literally within weeks. The results will be devastating."*

Julian Snyder makes a similar statement:

> *"The stock market is doomed by the dollar and a faltering 49 U.S. economy... The Swiss and Germans*

> *will sell dollars mercilessly the way they did in 1978 and Japan will stop buying U.S. bonds, perhaps delivering a financial Pearl Harbor."* - Julian Snyder, *The Armageddon Letter*.

The American job drain includes skilled and managerial positions. Computer makers are relocating all sorts of jobs to low-wage Malaysia. "There are some new high tech jobs, but they are not reserved for you in the First World", says Edith Holleman of the House Committee on Science, Space and Technology. "As international corporations move their facilities to cheaper locations, jobs in fields such as product design, process engineering, and software development are moving with them."

Mexico is not the only low wage area in the world by a long shot. Malaysia is a perfect model of what the New World Order people want us all to be like. They are docile people who quietly do what they are told while working for measly wages. American corporations are starting to invest billions in India. Already India is producing a staggering amount of software for computers, which has been one of bright spots in the U.S. economy. Next in line are cars. Nearly every car maker is thinking about building plants in India. The average per capita income is about $350 per year, far less than in Mexico. "Free" trade makes it very attractive for these companies to abandon American plants, which is what is really behind NAFTA and GATT. Check which politicians voted for them and where their political contributions come from.

GATT AND NAFTA

> *"The underhanded way Dole and Gingrich forced the 20,000-page GATT agreement down our throats illustrates their ethics. Do not expect them to champion any changes that will significantly benefit the American people."* - *Monetary Digest*

Why were Dole and Gingrich in such a hurry to get GATT passed before the new Republican Congress was sworn in? In questioning their motives, Patrick Buchanan said:

> *"Since the GATT treaty does not have to be ratified until next summer, why are these people so terrified of 45 days of hearings? Answer: They are afraid the American people will wake up to what is at stake in GATT. And what is at stake is nothing less than the destiny of their country. In the World Trade Organization (WTO) established by GATT, America surrenders her national sovereignty, her freedom of action to defend her own economic vital interests from the job pillagers of Tokyo and Beijing. We give up our freedom - to foreign bureaucrats who will assume authority over America's commerce that the Founding Fathers gave exclusively to the Congress... And, if we are outraged by the WTO's decisions, we have just one vote, out of 123, to challenge those decisions. The Europeans, who dreamed up WTO, have thirty votes. And in WTO, the U.S has no veto power.*
>
> *"Why would the most powerful country on Earth surrender such power to a foreign-dominated institution headed by a Third World bureaucrat? Because the politicians who negotiate these treaties, and the Corporate Elite who slave for them, have a different agenda than the national interest of the USA. They want to move America into a New World Order where the World Court decides quarrels between nations... In that Brave New World of their imaginings, the might and power of America are gradually transferred under global control."* - Patrick Buchanan

Dole, Gingrich, and the others pushing to pass NAFTA and GATT knew or certainly should have known about the inevitable devaluation of the peso shortly after the Mexican elections. This made Mexican labor twice as cheap as it already was. NAFTA would have never passed if the devaluation had happened before the vote.

American companies are firing Mexican workers who attempt to form unions. The multinational corporations are determined to keep low wage, free entry production on our doorstep. According to all of the hype, NAFTA was supposed to dramatically increase business for us from all of these newly empowered Mexican workers. It doesn't work that way and anyone with any common sense and understanding of economics would know that. With all the production moving to Mexico, why would the Mexicans buy more expensive products from the U.S.?

The U.S. border cities are a good example. Why would Mexicans cross the border to shop at a U.S. Wal-Mart, when they can shop at home more cheaply, because Wal-Mart only pays its clerks $5.50 a day in Mexico? Why would they buy American cars in the U.S. when GM, Ford and Chrysler pay their auto workers about $2 an hour in Mexico so the cars cost less there. GM has fifty assembly and parts plants in Mexico. Ford and Chrysler don't have as many, but Chrysler, which was a very strong backer of NAFTA, is planning on building more. Where are all the customers for American products in Mexico? Those big Wal-Mart stores are virtually empty. Who sold us this phony bill of goods?

Mexico's GDP dropped ten percent from the fourth quarter of 1994 to the first quarter of 1995. It dropped twenty-five percent from the first quarter to the second. The trend is accelerating. U.S. exports to Mexico have plunged. So much for all those American jobs NAFTA was going to create.

Other Latin American countries are jumping on the bandwagon as well. They are making deals with Mexico or Mexico and the U.S. to join in the duty free bonanza at our expense. The official propaganda is that they want to have cheaper access to our markets for their products and that they will buy more of our products if there is no duty on them. That is true to a certain extent. However, what they really want is the investment capital and jobs from all those U.S. manufacturers who will build plants in their countries and export U.S. jobs there as is happening in Mexico, Malaysia, and other countries. Then, they can buy U.S. products made in their countries more cheaply than they can import them from the U.S.

The government now admits that NAFTA will result in an estimated $2.5 billion in lost tax and duty revenues. That is a drop in the bucket compared to lost business and jobs. The latest estimate is that by July 1995, NAFTA had already cost one million jobs in the U.S., and we have hardly started shipping jobs south. Wait until all those factories get built in Mexico and the other Latin American nations join NAFTA and start importing our jobs.

According to Thomas Donahue of the AFL-CIO:

> *"You have to cut through all the fuzzy rhetoric and corporate propaganda to get to what NAFTA is about. U.S. manufacturing workers make $16.17 an hour (average salary and benefits) and have protections such as Workers' Compensation and environmental laws. Mexican manufacturing workers average just – $2.35 per hour, and only $1.64 an hour in U.S. dominated maquiladora plants. Hundreds of thousands of U.S. working people would lose their jobs, and NAFTA would put downward pressure on the wages of millions more."*

Kennedy Gammage has a similar opinion of NAFTA:

> *"The absolute truth about NAFTA: it is, will continue to be, a disaster for the American working man, despite the benefits to the big American corporations controlled by the insider elitists, who also control both parties in Congress as well as the media ... the dear old Eastern Establishment, i.e., Trilateral Commission /Council on Foreign Relations crowd.*
>
> *"Don't you believe for a minute the blarney about 'free trade'. There's nothing free' about it. You're talking about managed trade-mercantilism, or more aptly, a sort of multi-national corporation socialism/fascism."* - Kennedy Gammage, *The Richland Report*, December 30, 1994

Mr. Gammage goes on to say the GATT is much worse than NAFTA. GATT is just more of the same and includes a lot of other countries. It is the next step in the plans of the Trilateral Commission

It's The Stupid Economy - 83

to set up a one-world government, step by step, by entangling trade alliances with unelected, faceless bureaucrats from other countries replacing national sovereignty.

According to British Financier Sir James Goldsmith:

"The civil servants loved it (GATT) because they were going to be given all of the power; the politicians were getting rid of yet more responsibility but keeping the privilege. There was a general, almost conspiracy, wall-to-wall, of the Establishment against the people."

NAFTA and GATT are just more holes in the bottom of our financial boat. Our enormous debt will soon spiral out of control, due to compound interest and continued deficits. We owe more money than the external debt of all the other countries in the world put together.

"A new kind of financial panic is beginning. It's unlike anything we've ever seen - a stampede by large investors to unload billions in high-risk investments, with virtually no control by any government agency - no bailout mechanisms or fail-safe devices. I call it the 'derivatives panic'." - Martin Weiss, *Safe Money Report*, December 7, 1994

"We are at the same point as 1987. All year stocks had been moving higher even though the economy was near the breaking point. Then in August of that year, the Fed did the unthinkable. They raised interest rates. It was the kiss of death. The market, jittery over sky-high stock prices, began to turn downward. Less than three months later, Black Monday struck. The 1929 crash was almost an exact carbon copy: weak economy, bloated stock market, and then rising interest rates bringing on a collapse. Well, the Fed has just done it again. On February 1, 1995 – while our economy was struggling

> *out of its worst downturn in eight years-they raised interest rates!"* - Nicholas Guarino, *The Wall Street Underground*
>
> *"The carnage in the bond market in 1994 was one of the worst financial disasters of all time ... bonds were relentless pounded, month after sickening month. Investors in 30-year treasury bonds lost nearly 24% of their principal, as compared to 22% lost in the Dow on Black Monday, 1987. The bond debacle struck almost everywhere. Investors lost about $500 billion in Treasuries, $300 billion in corporate bonds, $200 billion in munis and another $500 billion in mortgages and mortgage bonds. Grand total: About $1.5 trillion in losses. That's significantly more than the losses of any previous market debacle."* - Martin Weiss, *Safe Money Report*

The Federal reserve has raised interest rates about 2.5% and investors have lost over a *trillion* dollars in bonds. Despite the recent reprieve with the upswing in long term bonds, the enormous derivatives market still hasn't been able to unwind to any great extent. The defaults in Texas and Orange County are just the beginning.

People are getting jittery with good reason. New car sales are down. New home purchases are down. All the things that go with new homes, such as carpets, are down. Short-term interest rates are nearly equal to long term rates, although that has eased a little with the recent small reduction in short term rates. If the yield curve does invert (short term higher than long term), look out. This always signals a serious recession.

> *"The Fed created the conditions for the derivative frenzy to take place. It intentionally made it possible for banks to make billions of dollars of profits to restore the strength of their financial statements. Yes, it is estimated that the banks made about $160 billion in profits this*

> *way. This was at the expense of savers who saw their returns on savings plummet by more than 50%. The Fed created a situation where the money went from the poor retired folks, depending on fixed income for their livelihoods to the banks that needed stronger balance sheets."* - Bert Dohmen, *Wellington Letter,* January 1995

Despite the shift in money from you to the large banks, the banks are still in trouble. That's why Clinton illegally bypassed Congress with his unconstitutional executive orders and gave Mexico $20 billion of your tax money from a slush fund that isn't legal either. He wasn't bailing out Mexican workers as he claimed, he was bailing out the banks and brokerage houses who had invested so much of their own and also their client's money in Mexico. The American taxpayer subsidizes the banks' bad investments again.

> *"Another point, the Republican's victory and their speeches about cutting taxes and big government have created a degree of optimism in the markets. Investment prices presently reflect the expectation of good times ahead. But you know my feeling. Republican politicians are every bit as power hungry and crooked as Democrats, they just sound different, and when the markets adjust for this reality, look out. Add it all up and it says that if the Fed does not soon reverse course and begin driving interest rates down, a big shakeout can't be far away."* - Richard Maybury's Early Warning Report

The Fed has recently dropped interest rates slightly because the signs of the coming recession/depression are becoming more obvious. Lowering the interest rate makes our bonds and dollars less attractive. Right now the manipulation of the dollar by the central banks is pushing it higher. When that falters, look out.

We are in a period of world-wide deflation. Japan's stock market and real estate bubbles have burst. They are at about half the value they were. Eventually they will go still lower. Recently Japan started re-inflating their money supply. That means they are creating money out of thin air to try to stimulate their economy and help lower their export prices. They are trying to reduce the value of their money, just as we did some years ago. Apparently, our disaster didn't teach them anything. They are using those new yen to buy gold and U.S. dollars. Gold is incredibly cheap in yen terms and the Japanese people know that as more yen are created, their monetary assets will go down. So, the people are buying gold as a hedge against Japan's deliberate inflation. Even with this re-inflating, Japan's banks, which are the world's largest, are in serious trouble. If even one of the top banks fails, watch out. The whole global banking system could collapse, as it is a house of cards.

Germany has its own problems including one of the world's highest labor costs. Why do we care? Guess who had been buying most of our debt – Japan and Germany. Japan in particular has been drawing back from doing so recently. As this trend continues, our interest rates will skyrocket devastating both the bond market and the stock market.

The stock market is so overvalued that it must enter a severe bear market or even collapse. It is in the final stages of a blowoff. Eventually it always returns to fair value and it will this time probably before March, 1996. Usually there is a serious overshoot in the correction, taking it well below the point of fair value. No one who is involved in our 1995 version of the tulip mania will believe this, but I hope that you remember the following. I predict that the Dow industrial average will eventually fall below 1,000! How much of your assets are at risk? Do you have a company or state retirement account? Do you have IRAs or 401s or whatever? Are you invested in a stock or bond mutual fund? If you have any control over where they are invested, *get out of stocks and bonds*

Cash will be king for a while until the Fed starts creating money out of thin air by the billions to pay the government's soaring debt. Then precious metals will go through the roof as our paper currency becomes virtually worthless. Gold and silver may well shoot up even before the Fed starts inflating like mad, as people the world over start ditching our worthless paper and buying real money. This is what is known as the flight to quality.

Most of the go-go mutual fund managers have never lived through a real bear market. They seem to think this bull market will go up forever and are staying fully invested in the market. When the market hits the skids, as it always does, they will panic. As investors try to cash out, they will be forced to sell huge blocks of stocks at any price they can get. Who will they sell them to? Much of the stock trading now takes place directly between funds. The stocks owned by the funds dwarf those owned by individuals. If they are all trying to sell stocks at the same time, who will buy them? It will be like a thousand people trying to go through a single door with a fire nipping at their heels.

> *"Few investors are aware that mutual funds can request permission from the SEC to temporarily shut down withdrawals (read the fine print in your prospectus)."* - Jim Stack, *InvesTech Mutual Fund Advisor*, January 13, 1995

I might add that they just might not answer their phones as happened in 1987. We could see an initial drop in the Dow of as much as a thousand points in one day. So far, these managers have looked at any downward correction as a buying opportunity and have jumped in to take advantage of the "bargains". When people start taking more money out of these funds than is coming in, watch out. Most of our meager savings is now invested in the stock market. When it blows, millions of people will have lost what little reserves they have and the economy will crater causing the loss of millions of jobs. Private household liquidity is already at its lowest level in more than thirty years.

> *"The present liquidity cycle shows that liquidity in the U.S. is tight and disappearing fast. It is thus not the time for equities. The liquidity cycle...is moving into the area in which commodities and other real assets are the best buy."* - Dr. Christofer Y. Thomas, *Imtrac*, January 1995

> *"The next ten years will witness years of volatility equal or greater than that during 1931, 1932, and 1933, and shake investors out of their carefree dream world for many decades."* - Robert Precther, *The Elliott Wave Theorist*, January 16, 1995

> *"Many economists think 1995 could see debt and money crises across the world. Some trouble spots in the currency market: Argentina – Inflation is low, but investors are fleeing. To keep interest rates high and preserve the peso's hard link to the dollar, President Menem may trigger a recession. Canada – To keep worries about massive deficits from triggering a currency rout, the Bank of Canada will have to keep rates high. Italy – The lira is sliding amid concerns over a $1.2 trillion national debt and political chaos. Some think a debt default is possible. France – High unemployment and rising budget deficits are weakening the franc's ties to the German mark. Mexico – The peso continues to slide as investors turn thumbs-down to government's rescue plan. It may take $40 billion in international aid to help. Spain – Central bank is raising rates as political concerns push the peseta to a record low against the German mark. More depreciation is likely."* - William Glasgall, *Business Week*, January 16, 1995

> *"This has been the longest period of negative money supply growth since the Fed was established in 1913, and the weakest money supply growth since the Great Depression ... with annual M3 money supply growth falling at an accelerating pace since May 1994, a 'new' recession of a new down-leg in a protracted great contraction ...is likely in early 1995. Liquidity problems tied to the dollar and the banking system continue."* - Al Sindlinger.

Currently, we are being kept afloat by Eurodollar borrowing at more than $200 billion a year and bond sales to foreign banks. Japan's re-inflation is helping, deliberately, to drive up the value of the dollar for now, but that won't last long. In order to keep attracting foreign money to a chronic spendthrift who can't pay its debts, we will have to raise interest rates. When that happens, the bond market and then the stock market will drop like a rock. We have managed to put off the day of reckoning so far this year, thanks largely to Japan and the British buying our junk paper, but it can't last.

Japan's economy is approaching the crisis point. The largest credit union and one of the largest banks have been closed. With a deflating economy, finding buyers for the assets will be difficult. Property values will slump. Car exports have dropped dramatically the last few months. Japan's problems may well be the trigger for a worldwide recession or depression.

Is there any way for us to avoid the coming depression? Yes, but we won't do what is necessary. In fact, both the Fed and the government will do all of the wrong things when it starts happening. We need to go back on a gold standard. That is what will have to happen eventually when people lose all faith in worthless paper money. We need to drastically cut government funding. Will we? No, not with the people now in office. Will we pay a horrendous price in the loss of assets and freedom? Yes!

> *"If we had a gold-based monetary system today, with its attendant far lower interest rates, we would save well over $100 billion a year in interest charges on the national debt. Compounded over five years, that's a net savings of almost $700 billion. Even in Washington, that's real money. Isn't it time we overrode economists' irrational phobia against gold?"* - Malcolm S. Forbes Jr., *Forbes Magazine,* January 16, 1995

> *"A sound monetary policy is the key to a prosperous economy, but the Republicans have no interest in monetary reform. Recall that it was a Republican President who sealed the fate of the gold standard, giving us two decades of rapidly declining purchasing power."* So it is today. Jim Leach, the new chairman of the House Banking Committee, praises the Fed (Federal Reserve) as a great institution. Alfonso D'Amato, Chairman of the Senate Banking Committee, loves Alan Greenspan

> *... In general, Republicans want a more powerful and secretive Fed"* - Former Congressman Ron Paul.

What can you as an individual do to lessen the impact of the coming economic collapse on your family? First, get out of the stock and bond markets if you are in them in *any* way. Then, get out of debt as much as possible. Develop a second source of income. (See Chapter Fourteen.) On a more collective level, help elect people to office at all levels of government that believe in the constitutional limits on the federal government that have long been ignored and also believe in an honest money system based on real money. Can we find and elect such people in time? I doubt it, but we have to eventually, anyway, or end up in a dictatorship. Let's start today. (See chapter Thirteen)

SOCIAL (IN)SECURITY

> *"If we can prevent the government from wasting the labors of the people, under the pretense of taking care of them, they must become happy."* - Thomas Jefferson

The greatest Ponzi scheme ever dreamed up is our so-called Social Security system. If you are not familiar with the term Ponzi, it refers to a pyramid arrangement where those who get in early get the benefits from the money contributed later by people who get left holding the bag. The scheme requires an ever-expanding base of suckers and must eventually collapse, even if it is run by a government. That is one of the reasons more and more governments are trying to back off from their socialistic programs. They are running out of more suckers/taxes to come in and pay for them.

The very structure guarantees the eventual collapse of any such scheme. That is why Ponzi schemes are completely illegal, except those run by the government. That is also why some countries have now privatized their Social Security system and are requiring that the money contributed actually be invested in something that provides real returns, so the system can be made actuarially sound. Not our government. To the contrary, it is raiding the funds to cover up how bad the deficit really is.

Several things are going to make the promised benefits unavailable in twenty years or less. First, all of the money that has been collected from you isn't there. The government spent it all to pay for all of their pork barrel projects which is a euphemism for buying people's votes. The treasury did give the Social Security Administration "notes" for the money they swiped but these notes cannot be sold on the open market. Only the treasury can redeem them, *if* they can *borrow* the money somewhere. You paid money in for your retirement; the government spent it. The Treasury doesn't have any money in your "trust" funds. They don't have it now and things will only get worse in the future as interest rates rise, the dollar falls, and nobody wants to loan the government more money to throw away.

> *"It is a fraud to borrow what we are unable to pay."* - Publius Syrus

> *"Today's politicians don't pass the buck as much as they spend it – and steal it."* - Tom Anderson

Even if the money was actually there, the whole scheme still wouldn't work. First, the whole system was formulated on retirement at age sixty-five at a time when only a few people were lucky enough to live to collect it, and they didn't live very long afterwards. With the advances in health care and the graying of America, the system cannot hope to keep its promises. On top of that, thirty million babies have been murdered, some of whom would now be entering the work force to start contributing to the support of those retiring. They aren't there. In the early days of Social Security, there were twenty people working and contributing for each one drawing benefits. Now the ratio is down to three to one, and soon every two working people will be supporting each retiree.

The predictable revolt of the younger people has already started, with organizations being formed to stop/reduce the benefits, and who can blame them? Between social security and Medicare costs, euthanasia (as in rationed health care directed to the more productive members of society) is right around the corner.

These are the choices facing the government:
1. Continuously reduce the benefits and advance the age requirements.
2. Privatize the system before it is too late (No Way!)
3. Print money, causing the destruction of what's left of the dollar's value.
4. Tax workers much more heavily.
5. Default on the debt.

Number 2 is out because the Congress will never give up access to that loot. Number 5 is not feasible, because the whole government would break down. So the government will use a combination of 1, 3, and 4, which will so burden the economy that we will quickly become a second class economy, making it even more impossible to meet the promises. In other words, forget about any real benefits for anyone trying to collect anything of real value after the year 2000.

What can you do about it? Well you could support only candidates that will stop the pilfering and take the system private or quasi-private. Do not count on Social Security and beef up your own retirement as much as possible. The problem with that is that Congress already plans to "tax" (confiscate) fifteen percent of your existing pension funds and then tax another fifteen percent each year. On top of that, they plan to reduce any Social Security you might get by what you get from other retirement sources, because you don't "need" it. Strange, I thought that Social Security was "insurance" in "trust" funds that we paid into and that we had a right to a return. Hmmm. Could it be that someone was fooling us all of the time?

Your best bet is to start your own business on the side that can still support you without too much work when you get to retirement. Most of the people now under fifty will never actually get to retire. The government loudly proclaims otherwise. They are lying and they know it. They have to claim otherwise to stay in power. Take their own numbers and run them yourself. It doesn't work. Even if the money was actually there, it still wouldn't work. Even with the most optimistic forecasts for the economy, which is actually moving in the other direction and will sink lower than any of us can imagine, it doesn't work. Try to become as self-sufficient as possible. Protect what you can. The government is coming to help you and take care of you.

In 1994, the Bipartisan Commission on Entitlement and Tax Reform released a report which said that by 2012, entitlements – including Social Security, Medicare, Medicaid, and civil service pensions – would consume almost all U.S. tax revenues. In a recent survey, twice as many young adults indicated faith in UFOs as expressed belief that Social Security would exist when they retire. They are probably right about the Social Security system.

Forbes Magazine (11/21/94) carried an article entitled: "The Big Black Hole" which was subtitled: "Social Security is more than $6 trillion in the hole. Its unfunded liabilities actually exceed the national debt." Forbes wrote:

> *"In present value terms, Social Security is $2.7 trillion in the hole with respect to current retirees. Social Security actuaries estimate that we would have to pay a further $4 trillion to settle claims of everyone who has paid into the system, but has yet to receive benefits. Think of the $6.7 trillion total as the amount the government would have to pay a private company to take over the current obligations of the system. And that $6.7 trillion total does not even include the government's unfunded Medicare liabilities."*

At the end of 1992, the Social Security Trust Fund contained assets of $331 billion – just over a year's worth of benefits. But these assets are all government bonds – IOUs from the rest of the federal government. The actual cash surplus from taxpayers that exceeded the amounts needed to pay for current benefits was spent by the government for general fund purposes (i.e., to cover or cover up part of the federal budget deficits) and the bonds which would have otherwise been sold to the public to pay for that federal spending were instead deposited in the Social Security Trust Fund.

As Tom Donlan wrote in *Don't Count On It*:

> *"Since the Social Security program began, Americans have been told that they are not really paying a tax, but*

> *rather making contributions to a retirement program, or paying insurance premiums. They have been told their money is in a trust fund, and that they will be withdrawing their money when at last they retire and begin to collect Social Security. It's a lie. It always has been a lie. From the day they paid the first check, Social Security has been paying today's beneficiaries with the contributions of tomorrow's retirees."*

So the Social Security system is actually insolvent. The trust fund has been plundered by Congress and several administrations since the mid-1980s. If there will be no money left to pay retirees in ten to twenty years, what is the government going to do to wiggle out of the situation? In 1993, it began to sharply increase taxes on Social Security payments received by retirees. There is now a growing move to push the retirement age up from sixty-five to seventy before a retiree can begin to receive payments. For retirees who only live to seventy-five, this will cut their Social Security benefits in half. If they live to eighty, the benefits will be cut then by one-third.

And now, according to Forbes, the politicians are considering phasing out eighty five percent of benefits for people with retirement incomes between $40,000 and $120,000. As Forbes wrote, "People obtain that kind of retirement security only by having saved during their working years through IRAs, 401(K)s or simply by putting money aside in a bank. Rewarding such thrift by taking away their Social Security benefits sends a devastating signal to the nation's savers." It is also a breach of contract.

As Don McAlvany aptly pointed out:

> *"As the Social Security crisis worsens, the government can be expected to rip off the people, to cut benefits to people who have paid them in, to tax beneficiaries, and to cover their own profligacy and malfeasance by further stealing from the people. Obviously, government promises or contracts with the people are not worth the*

> *paper they are written on. How can we have a contract with the government or plan ahead when they lie, steal, cheat, and continually change the rules for their benefit?"*

I would add that the government always reduces the benefits to the "rich" to protect the "poor" who really need it. This is another example of the government promoting class envy to try to cover up their reckless irresponsibility.

You had better do everything you can to protect your retirement assets. Our "official" national debt is over $4 trillion. That means that every man, woman and child owes somebody, mostly large banks, about $17,000. That is just the tip of the iceberg. With the unfunded liabilities, such as Social Security, the military, and government pensions, that figure is about $20 trillion. Then, there are all of the loan guarantees. The only source of money left without increasing tax rates that would devastate the economy is your retirement funds. Besides the tax mentioned above, the Democrats want to "free up" your money from "passive" stock and bond investments. They want to force you or your pension fund managers to invest in more "active" investments. These are known as either government bonds, or else just outright confiscation of your money as a "partnership tax" to fund the government's social income redistribution schemes. Maybe you had better get some of your retirement funds into gold and silver coins and bury them somewhere the government can't find them. You might also want to help elect people who have enough sense to drastically slash government spending to what is necessary to carry out the constitutionally allowed activities of the federal government.

> *"Sometimes the law defends plunder and participates in it. Thus the beneficieries are spared the shame and dangers that their acts would otherwise involve.*
>
> *"But how is this legal plunder to be identified? Quite simply.* ***See if the law takes from some persons what belongs to them and gives it to another person to whom it doesn't belong****. See if the law benefits one citizen at*

> *the expense of another by doing what the citizen himself cannot do without committing a crime.*
>
> *"Then abolish that law without delay. For it is not only an evil in itself but also a fertile source for further evils because it invites reprisals and imitation. If such a law – Which may not be an isolated case – is not abolished immediately, it will spread, multiply, and develop into a system. No legal plunder; this is the principle of justice, peace, order, stability, harmony and logic."* - Frederick Bastiat, *The Law*, 1850.

In other words, there should be no income tax to take money from those who earn it and give it away to those who don't in order to buy votes and build political empires.

THERE IS NO GOOD TAX, BUT...

> *"Noah must have taken into the ark two taxes, one male and one female. And did they multiply beautifully."* - Will Rogers

There is no such thing as a good tax. Any tax that takes money that might be used to purchase more goods which puts more people to work, or to invest it so that it may help someone produce more is a net loss to economic growth. Governments produce nothing except hassles and headaches. Unfortunately, both governments and taxes are necessary evils. The task of the common man is to reduce both to the bare minimum necessary to protect society. If I may paraphrase Lincoln, the best government is the least government. Our Constitution was written in order to provide for the common defense of all states, promote commerce between the various states, promote foreign trade, and keep order within the country.

Everything the Constitution *allowed* the federal government to do falls within those categories. We have gone far beyond those essential constitutional permissions. The chains that the authors of the Constitution thought they had applied to the central government to keep it in check have turned out to be rubber bands at best, thanks in part to Justice John Marshall.

One of those chains was thrown away when the Sixteenth Amendment was passed, allowing direct taxation of individuals. It never was actually ratified, but we have been stuck with it anyway. Let's look at some of the problems that were caused by removing that chain and then some solutions.

INCENTIVE

The Democrats are currently braying about tax cuts going (just) to the rich. They know better, at least most of them do, and those that don't shouldn't be in positions of power. However, it sounds good in their never ending promotion of class envy and their attempts to divide people into warring camps for their political benefit. When you suppress incentive, as their approach to taxation does, people just don't do the things that create more jobs, create new products, and increase everyone's standard of living. Many businesses and investors today spend more time and effort dealing with the tax consequences of their actions than concentrating on sound business principles. Many people who could start their own businesses and provide jobs for others, don't do so because of all the hassle, expertise needed, and time (money) wasted just dealing with all of the tax laws and other regulations. The punishment of everyone who succeeds in spite of all of the obstacles, by these "soak-the-productive" politicians, is a powerful disincentive to anyone looking at the risk and incredibly hard work that it takes to succeed in a new business. Since large corporations, particularly those in product production industries, are continually losing jobs and almost all the job creation is taking place in small businesses, these politicians are shooting each of us all in the foot to try to maintain their political base by promoting class envy and division.

Take a look at the following figures of who pays the income taxes in this country.

Top 1%	(income)	pays 26% of taxes
Top 20%		pays 75% of taxes
Bottom 60%		pays 10% of taxes
Bottom 40%		pays 1% of taxes

In other words, half the working population virtually gets a free ride. Then there is the non-working group, who get their way paid by those who do work. Yet, every time a tax reform proposal comes up to try to put some true fairness in the system, the Democrats scream that all the tax breaks are going to the "rich".

The Democrats say that cutting the capital gains tax amounts to "giving" (that's their word) money to the rich. There are two things wrong with this statement. First, the money belongs to the people who earned it to start with. Most politicians, particularly those in Washington and especially the Democrats, think that tax money belongs to the government to start with, and in reducing taxes, the government is "giving" too much of *its* money to people who don't deserve it, not because they didn't earn it, but because they don't "need" it. This is the basic philosophy of communism.

Let's look at this "need" business. A foundation principle of communism is, *"From each according to his ability; to each according to his need."* Anybody who knows anything about human nature knows that this is one of the silliest statements ever made. Like most liberal/socialist theories, it sounds good on the surface, but falls apart in practice because it ignores basic human nature. As a basic principle of an economic system, it dooms the system to failure as we have recently seen in the Soviet Union. I won't go into all of the details as it should be obvious from history, if not common sense, but let me point out that slaves have little incentive to produce anything, other than by force or coercion in some form.

Despite the obvious nonsense of "need"-based economics and the accompanying destruction of incentive, this is precisely what our politicians keep imposing on us. Every time you hear a politician say someone needs something (e.g. welfare, "free" medical care), or doesn't need something (confiscate what they have), remember the source of that philosophy is the very root of communist/socialist economic theory.

Many people and most politicians can't get past the "obvious" idea that if you raise taxes, you raise tax revenue. It may be "obvious", but it is dead wrong, particularly in the long run, if you are taking too much out of the economy. Let's look at a simple example. In earlier centuries when farmers couldn't just go to the feed store and buy their seed, a farmer had to save a certain amount of his crop to provide seed for the next planting. If he had to eat most of his crop to survive, and therefore couldn't save enough seed to plant a full crop the next time, he got an even smaller crop and had to eat even more of his seed to survive, until he eventually starved to death for lack of seed to plant. Money saved by individuals is the seed of a free capitalistic economy. It is the money invested by individuals in starting new businesses. It is the money borrowed by businesses to build new factories, buy more modern equipment, and create new jobs. Every time the rats (government) eat (tax) the seed, the crop is reduced. This is a cumulative effect. The time and crops lost can never be recovered, and we are forever poorer for it. Increasing taxes beyond a certain point, shrinks the economy, lowers the GDP from what it could have been, and reduces everybody's standard of living, except perhaps the rats (politicians and bureaucrats) who live off of the seed. They can get rather fat while the rest of us get skinnier.

The effective take home pay of the average wage earner has been declining since the 1970s. When I got out of college in the 1960s, a single wage earner family could usually afford to start out in a new home. Now a two wage earner family has to struggle to buy any kind of a house. Has somebody been eating our seed? Remember, in politics (or economics), nothing happens by accident. The U.S. has the lowest savings rate of the industrialized nations, thanks mostly to our tax policies. Japan has the highest. In 1960, a dollar "bought" 180 yen (or products). Today it buys about 80 yen. We used to produce most of the world's consumer electronics. Today, we produce virtually none. We used to have the four largest banks in the world. Now the six largest banks are Japanese. A few years ago we were the world's largest creditor nation. That means the world owed us money. Now we are the world's largest debtor nation. We owe far more than we can ever repay. Who borrowed all that money (the rats) and why (to stay in power)? We certainly did not need it to pay for the constitutional functions of government. The politicians borrowed it to keep the incumbents in office and advance socialist policies. "We", not the politicians who will retire on very fat pensions, will have to repay it, one way or another.

One of the reasons Alexander Hamilton argued so strongly for consumption taxes, that is taxes on trade, rather than income or property taxes, is that consumption taxes are much more obviously self-limiting. If you raise sales taxes, people can't buy as much, because the products cost more. You quickly price things out of the reach of people. The same thing happens with income taxes, but it is much more subtle, especially with business income taxes. As originally set up in the Constitution, the government could only tax a limited amount (on trade) without killing the goose, thus limiting government, which was exactly the idea. That's why the Constitution (originally) forbid direct taxes. All of that common sense and protection went down the drain with the Sixteenth Amendment, which allowed an income tax. Now look at where we are.

To make it a little more obvious what happens in regards to income taxes, lets look at the extremes. If the tax rate is zero percent, no taxes are collected and no tax income is produced. If the rate is 100 percent, there is no reason to work as everything is confiscated and you starve to death anyway. So, nothing is produced and no income to tax is generated. Also, there is no money (profit) left to buy new equipment or supplies to work with. Somewhere between these extremes is the point where you can collect the maximum income from taxes without hurting the economy enough to start reducing tax income. A curve of revenue versus rates has come to be know as the Laffer curve. The peak revenue production comes somewhere in the low twenty percent range. Counting all of the taxes we pay, many of us are in the forty percent range. Increasing taxes now results in lower tax income to the government eventually. It may take a while to show up, particularly in an expanding economy, but eventually taking the money out of the economy and pouring it down the government rathole takes its toll. Reducing taxes actually increases tax revenue at this point, as has been proven time and again. About seventy percent of our economy is consumer buying. If you take money away from people, they can't buy as much and the GDP (economy) shrinks. Is that so hard to understand? For some reason, most of our politicians just don't seem to get it.

So, what is the best tax plan? Well, there is no good tax, let alone a "best" one. Taxes are a necessary evil. The real object should be to minimize the evil by limiting government to the minimum necessary to provide reasonable security and facilitate trade. That is precisely what the framers of the Constitution did. This is exactly what "we" undid in 1913 and now, as a result, we are hopelessly in debt.

> *"The beginning of the end of capitalism in America, which was also the beginning of the rise of totalitarian communism, came in 1913." "This is why the decline of capitalism should be dated from 1913, when the 16th Amendment authorizing the federal government to impose an income tax was adopted."* - Joseph Schumpeter, Economist.

Income taxes pose three major problems. One is the illusion of seemingly unlimited income available to the government. That illusion promotes borrowing against future revenues if they can't tax you enough now. The second is the massive destruction of privacy required to collect it (IRS). The third problem is that there is no way to make taxes "fair", whatever that means. For these and other reasons, direct income taxes were forbidden by the writers of the Constitution. For the same reasons, we need to get rid of them. To have the massive and destructive redistribution of wealth and resources from those who are productive to those that are not, that the socialist and primarily the Democrats want, you have to have direct taxes, whether you call it an income tax or "contributions" to Social Security, which is just another tax.

> *"I place economy among the first and most important of republican virtues, and public debt as the greatest of dangers to be feared....And to preserve (our) independence, we must make our election between economy and liberty, or profusion and servitude."* - Thomas Jefferson

The only type of tax that doesn't directly punish accomplishment, break up family holdings or destroy incentive is sales tax. It is also the only type of tax that you have a choice of paying by deciding whether or not to buy things. There is no invasion of privacy with a sales tax. It is also much harder to cheat. Even drug smugglers buy planes, boats and cars. Now they pay no taxes on their illegal profits. One problem with sales taxes is that they are regressive. That is, they fall more heavily on the poorer people who have to spend all of their income to live reasonably. That objection can largely be eliminated by exempting food, drugs, food supplements, housing, and medical care from the tax, or at least having a much lower rate for these items.

Property taxes were also a no-no with the authors of the Constitution for good reasons. The power to tax property creates the power to confiscate it for the non-payment of those taxes. An obvious result of property and inheritance taxes has been the breakup of the family farm in America. All too often, inheritance and property taxes either singularly or combined, have forced the sale of farms that have been in particular families for generations. Older people have been forced to sell the homes they have lived in and worked for most of their lives because they could not afford the property taxes. Is this what we want from our government? It is what we have. As long as the government can impose property taxes, it can simply confiscate all property by continuing to raise the tax.

We need to overturn the Sixteenth Amendment because it was never properly ratified, or repeal it if the Supreme Court wimps out. If I were somehow magically appointed President, one of the first things I would do is throw the resources of the Justice Department into getting the Sixteenth Amendment thrown out, since it never did receive the necessary two-thirds vote. If someone outside of the government tried to get it overturned, the Supreme Court would probably abandon the written requirements of the Constitution and say that it has been accepted too long and thus it would wreck the government and the country to get rid of it. If the government itself attacked its ratification, it would be much harder for the court to justify that lame argument, but they probably would still do so.

We need to limit runaway government by limiting its income to that which is necessary to provide the essential constitutional functions. The flat rate tax idea is a step in the right direction, but it doesn't go far enough. There are still several problems with the flat tax concept. One is that the IRS would still invade our privacy. The flat tax should (but probably won't) reduce the IRS and its cost. Another problem I have with the flat tax is that a lot of people would be exempted from paying any tax at all. I think that everyone should pay something for several reasons. Why should you have to pay for my defense, or any other benefit I receive from Washington, just because you earn a little more money? Also, those who pay no taxes will gladly vote for candidates promising more benefits to them, since it costs them nothing but raises taxes for those who do earn enough to be taxed. That is part of what is wrong with our system now. The people who demand the most from the government pay little on nothing in taxes. Everyone should pay at least a little something so that they reap the same result when they vote for people who want to raise taxes or borrow more money, which is worse.

The whole idea of direct taxes was forbidden in the Constitution in the first place, and we need to go back to that philosophy. The experiment of direct taxes has been a colossal failure just as our forefathers knew it would be. The cost of compliance with our current tax chaos is about $600 billion. That's the direct cost. That does not include the taxes, just the compliance costs. Who knows what the total cost is? Can you imagine the result if that $600 billion was added back into the economy by getting rid of the income tax? Can you imagine how much more competitive our businesses could be without that cost burden?

According to the Tax Foundation, Americans now pay a total of $2 trillion a year in taxes – federal, state, local, Social Security, and sales taxes. The average American spends $3,800 a year for food, housing, and clothing, but he pays $7,927 in taxes. Taxes are the biggest factor in Americans' rising cost of living. Of course, to mask the effects, the government refuses to count taxes in the cost of living index. Clinton got the biggest tax increase in history. Quite predictably, the value of the dollar is falling and we face another recession, if not a depression in the next couple of years.

After throwing out the income tax, I propose the following tax structure. First, impose an import tax of no more than three percent. This low rate should not start any trade wars. The point here is income production, not protectionism. It would give a very slight advantage to American products and labor, but not remotely enough to overcome second or third-world wages to any extent. If some countries would retaliate, so what? Only twenty percent of our economy produces anything we can sell, and most of that is used internally. If we imposed this import tax, and other countries followed suit as a way to raise income, the playing field would still be basically level anyway.

The second tax I would propose is a national *retail* sales tax not to exceed ten percent. Any more would be clearly destructive to the economy. Wouldn't this reduce the government's income drastically? Exactly! It would also encourage savings and investment, both of which are desperately needed to compete in a global economy. Think about it: no forms to fill out; no records to keep, no IRS prying into your bank accounts or investments, no April 15th panic. While the total amount taken from you by the government is not obvious with a sales tax, it is rather easily figured if you want to know. More importantly, every time you buy something that is priced at $10 and you pay $11 for it, you feel the bite. You are much more likely to vote for people who will reduce the bite rather than increase it. Since sales taxes hit everybody, this aggravation would be almost universal, which is what we need to bring the government under control.

One of the sneaky things about income taxes and property taxes is that most people don't really realize what they are paying. These taxes are built into withholding and mortgage payments. What people look at is their take-home pay and monthly payments. Why not take home all of it and *you* decide if and where to spend it? Politicians love these taxes because people don't really realize what the government is costing them. With the sales tax the "taxing" is much more obvious and aggravating, which it should be. That way you will think about how you will vote or if you will vote next time and that's what the politicians hate.

> *"If taxpayers over the years had paid their taxes the same way they make rent or car payments, the government never could have grown as large as it has. Only by taking people's money before they ever see it has the government been able to raise taxes to their*

current level without igniting a rebellion. My (flat tax) plan would end withholding and put a permanent check on the confiscatory appetites of the political class." - Rep. Dick Armey, *Forbes Magazine*, January 2, 1995

In regard to sales taxes, you may have noticed I said *retail* sales taxes. We must avoid, at all cost, value added taxes. This is one the politicians really love, as you have no idea how much is going for taxes. It is also a bureaucratic and bookkeeping nightmare with a hopeless tangle of rules. If the sales tax is actually collected as a tax added at the final point of sale, then you see *your* money that *you* earned actually leaving *your* hand as a direct tax, and not disappearing before you get it or hidden in the ticket price of the item.

One of the greatest advantages to the above sales tax plan is that it eliminates most of the infighting over who gets what tax breaks that makes the income tax such a hopeless mess. There are no tax breaks to fight over. There is, therefore, no reason for many of the PACs that so corrupt our political system. There is only enough income to pay for essential constitutional government functions and no "spoils" to divide. If we had stayed with the original taxing powers in the Constitution, we would be the most prosperous nation in the world with the highest per capita income (no, we are not now). We would still be the world's largest creditor nation, far larger than we were, and we would not be bankrupt now, which we actually are. Some might argue that we owe so much that we can't go back. We can and we have to do so. If not, the present course will mean not only the destruction of the economy, but of our freedom as well. We have to bite the bullet sooner or later. The debts have to be paid. The only choice we have is how to do it, and the sooner the better, to minimize the pain.

If you think the Republicans are going to do anything to seriously address the mess, think again. According to Newt Gingrich: *"We're never, ever talking about cutting spending. We are talking about a slower rate of growth .... so, first of all, we're talking about less growth. We're never talking about cutting."* - (Meet the Press, December 4, 1994.)

Don't depend on Newt Gingrich to really change things very much. He's a Rockefeller Republican and a member of the CFR. It is no surprise that he has received so much press, far more than any other speaker, as he has been endorsed by the liberal media and the Establishment. He has supported most of the tax increases of the past six years. He believes in big government and has campaigned for both NAFTA and GATT. It appears that he is being groomed by the CFR/TLC people to be President if the country swings more to the right, While Colin Powell, who is also a member of the CFR, is being hyped as a "moderate" candidate if that stance seems to have more appeal.

> *"Well, it has begun: the Republicans' Contract with America But our hopes will not come true....The public wants to believe in quick fixes, painful only to the other guy. They vote accordingly. The politicians know this. Until the public is willing to roll back government over a 40-year period, the Contract with America is like a New Year's resolution: made to be broken."* - Dr. Gary North, *Remnant Review,* January 5, 1995

> *"These problems are greater than anything the Gingrich program addresses. Even if the new Congress were completely sincere in its reform efforts, I would remain skeptical, since correcting the ills generated by the Welfare State, an inflationary monetary system, and unpayable debt accumulated over the past sixty years, cannot be done without serious economic and political turmoil that can be expected and the increasing popularity of the militias, the 10th Amendment Movement, the underground economy, emigration, and moving wealth offshore, tells us that a significant number of Americans already know how serious our problems are. The recent elections confirm this attitude as well."*
> - *The Ron Paul Survival Report,* January 15, 1995

"Tax and spend is not just an economic issue. The more government takes from your wallet, the more it takes from your Bill of Rights down the line ... Once the government has your money, the dollars inevitably drift toward the left and toward the secular. Left, because government action is inherently collectivist, and government bureaucrats exist to collectivize. Secular, because the Constitution excludes government from the realm of the sacred. The more the government sphere expands, the more the sacred contracts. It's as simple as that." - Peter Huber, *Forbes Magazine*, January 16, 1995

As if the U.S. tax system wasn't strangling us enough, now the U.N. wants to tax us also. The head of the U.N. has proposed an income tax of one percent of every citizen's income in the *industrial* nations. This money is supposed to be a safety net for the poor, as usual. In other words, they want to take your tax money and reproduce our disastrous welfare system all over the globe. Guess who will end up with most of that money, just like they do here? The bureaucrats. Actually, much of that money will go to support the U.N. armed forces to enforce "cooperation" with U.N. policies of a one-world government.

While one percent sounds like a very small sum, that is just how the U.S. income tax started out, just a minor amount with lots of promises about it never being significant.

The only way to limit government and government interference (control) in our lives is to put them on a strict budget that Congress can't get around. The balanced budget amendment might be a small help, but Congress will get around it, as they always do with any attempt to limit spending. The only way to limit government is to limit its power to tax and borrow. We must repeal the Sixteenth Amendment *and* pass an amendment forbidding the federal government from borrowing except in the time of war. Not until the government quits confiscating so much of the economy will our real income reverse its downward trend. The large gains in productivity in the past will be impossible to reproduce in a highly competitive global economy.

While large corporations are downsizing and eliminating jobs, most new jobs are coming from small business. Yet the government provides powerful disincentives to starting a business. The tax code is impossible to understand, let alone comply with. No one knows what all the regulations are for most businesses, so how do you comply with them? If delivering a Big Mac is covered by 40,000 regulations, what business is not hopelessly buried in government red tape? You may know all about how to produce a super widget, but you can't also be an expert on all the government red tape. So if you want to even try to comply with the most obvious stuff, you need to hire lawyers, accountants, and consultants. How do you do that when starting on a shoestring as many people do? Most businesses fail because of a lack of sufficient startup capital. Thanks to our government, much of that capital is wasted trying to deal with government regulations or even finding out what they are.

If we are to recover from the coming economic fiasco and provide jobs for people, we need to get rid of most of the disincentives to starting businesses that government requirements currently impose. We will also have to have an honest money system as mentioned before.

Chapter Four
Education or Indoctrination

"The aim of totalitarian education has never been to instill convictions but to destroy the capacity to form any." - Hannah Arendt

Have you driven past any red dinosaurs lately? I don't mean purple ones; I mean red brick ones. They are still called school buildings, but they are really dinosaurs. They are obsolete. They are the left-over remnants of a failed approach to education that we have been stuck with for over one hundred years. Not only has it failed, but it has become ridiculously expensive with most of the money going to the bureaucracy and administration, rather than the students. The goal of the educational system has become not education, but indoctrination in political philosophies, especially a particular world view.

I hope you won't mind a personal story. (Even if you do, it's my book.) A number of years ago, I was the executive vice-president of a computer company. We provided desktop computers and software to the government to train astronauts, submarine crews, pilots, and various other groups. We built the computers from scratch. This was before the IBM PC came out. We created the authoring software and designed and coded the training courseware.

I was looking for a way to expand our market and ran across a request for a proposal by a southern school system. They wanted a group of networked computers for both grade schools and high schools to test computer-based education. This program was not intended to teach students how to use computers, but to present actual class lessons. We had absolutely nothing that was suitable and there was no time available to write the courseware from scratch. Our authoring language, however, was very similar to that used by the Plato system running on mainframes at the nearby University of Illinois. I contacted the University and they agreed to let us use and modify a number of programs to run on our computers.

These programs contained nearly all college level courses. Some of them were remedial courses used at the college level. We tried to match what was available to the various grades. We did remove some of the hardest material and replace some of the words to try to match the reading/vocabulary levels of the appropriate grade levels as much as we could in the time available, but this was still mostly college level material. There were a couple of demonstration lessons for the earlier grades that we incorporated into the curriculum. I knew that we had a superior system to anything else on the market, but I also knew that ours would be much more expensive than the ones we were bidding against. Since this was a demonstration contract, I configured it as a lease with a buyout option to get the price down. Ours was still the most expensive, but it had some decided advantages. Fortunately, I was able to convince the school board to accept our bid.

At this time, none of the kids had even seen a computer, if you can imagine that today. In order to introduce them to computers and try to get them enthused about using them, I wanted to have a game that they could easily play even at the third-grade level. The only thing I could find on Plato was the old game of hangman to teach spelling and vocabulary. It was a typical instance of just using a computer to replace a piece of paper, and did virtually nothing to use the power and capabilities of the computer. There were two things wrong with good old hangman. One, it was boring. Two, I didn't think the image of hanging people was a very good idea for a southern school system that was mostly black.

I asked my best programmer if he could modify the graphics code to provide flying saucers and phasers instead of the gallows. I outlined what I wanted: the bad guys in the space ships attacking earth and the good guys on the ground. I inserted a three second delay and a lot of shooting between the choice of a letter in the game and the destruction of either a saucer or a ground installation, depending on whether or not the letter was correct to add a little suspense.

It was amazing what he came up with, particularly with the little Z-80 computer chip we were using. I haven't seen anything as smooth and as fast with as much action on a computer to this day.

Education or Indoctrination - 111

Anyway, we got everything together and shipped it all to the school system. The teachers and principals took a look at it and said it was a great system, but the kids could never handle the levels of material we had provided. Little did they know. We suggested they just give the kids a chance and see what happened.

What happened was that they couldn't get the kids off the computers. Not just the game, but the lessons as well. These kids, like all kids, really wanted to learn. We had just made it interesting and challenging. The computers were much more interactive than listening to a teacher drone on interminably. The kids became completely engrossed in what they were doing. They had to pass a test on each lesson before going on to see what was next, which provided an incentive to master the material. Like young eagles when they are kicked out of the nest, the kids flew – high. At the end of the year, the ones who had used the computers scored far beyond the other kids, not just on the computer lessons, but in all categories. School had become fun. Learning had become an adventure, not a boring chore.

The teachers loved our system. The principals loved our system. The superintendent loved the system. Typically, the school board cancelled their program to computerize the schools and used the money to hire more administrators and give the teachers a raise. The teachers who had worked with the computers would much rather have kept them and forgotten about the raise. The administrators won out over the children again.

I told this story to make a point. Our whole approach to education is largely obsolete. We have available a large group of teaching devices, including computers, videotapes, interactive disks, the Internet, and an expanding range of technology. While these are being used to a very limited extent, their potential hasn't hardly been touched. The present system of isolating kids by ages, rather than interests and abilities, placing them in rows of hard uncomfortable chairs, and trying to enforce an arbitrary level of instruction for everyone has always been absurd. It is the most inefficient method anyone could design. It is the easiest on the teachers and administrators and very counterproductive for the students.

John Taylor Gatto won an award for being Teacher of the Year in New York City in 1990. He has written a scathing denunciation of the present approach to education, with plenty of information to back it up. If you are interested in your children's education or in improving the educational system, you need to read his book, Dumbing Us Down. He lists seven things that the present system is designed to teach to students. They are:

1. Confusion.
2. Class position
3. Indifference.
4. Emotional Dependency.
5. Intellectual Dependency.
6. Provisional Self-Esteem.
7. One Can't Hide.

All of these are highly destructive to real education. You should read his explanations of them if you really want to know what your tax money is being wasted on and how most schools are indoctrinating your kids.

John's students were consistently superior to others in the same grades and system. Given that these kids were from Harlem and Spanish Harlem, their superior performance wasn't due to any advantages they might have had. How did he accomplish this miracle of modern education? It was by going back to many of the ways we used to educate our children before modern compulsory education. He circumvented the system as much as possible, breaking the rules, making learning fun, and teaching practical, useful subjects. He got them involved in doing things, not just sitting and listening. As much as he could, he let the kids study what they wanted to at their own pace. Guess what? It worked. Boy, did it work. That's why he won the awards and now speaks all over the country to people who are desperate to fix a badly broken and bloated system.

I really want you to read his book. Until you do, here are a few quotes from teacher of the year Gatto.

> *"...the truth is that reading, writing and arithmetic only take about one hundred hours to transmit as long as the audience is eager and willing. The trick is to wait until someone asks and then move fast while the mood is on.....Pick up a fifth-grade math or rhetoric textbook*

from 1850 and you'll see that the texts were pitched then on what today would be considered college level. The continuing cry for "basic skills" practice is a smoke screen behind which schools preempt the time of children for twelve years and teach them the seven lessons I've just described to you."

"School as it was built is an essential support system for a model of social engineering that condemns most people to be subordinate stones in a pyramid that narrows as it ascends to a terminal of control. School is an artifice that makes such a pyramidical social order seem inevitable, although such a premise is a fundamental betrayal of the American Revolution."

*"The current debate about whether we should have a national curriculum is phony. We already have a national curriculum locked up in the seven lessons I have just outlined. Such a curriculum produces physical, moral, and intellectual paralysis, and no curriculum of content will be sufficient to reverse its hideous effects. ......
Schools teach exactly what they are intended to teach and they do it well: how to be a good Egyptian and remain in your place in the pyramid."*

Many other people realize just how absurd our current mode of "education" is and what it really should be. At the risk of boring you with quotes, consider the following.

"The whole art of teaching is only the art of awakening the natural curiosity of young minds for the purpose of satisfying it afterward." - Anatole France.

I don't think that teachers need to worry about satisfying this curiosity. Just motivate the children properly and you will have a hard time stopping them. My experience with computer based education clearly indicates what can happen when leaning is made easy and fun for kids. Their thirst for knowledge cannot be quenched.

> *"Education is not to reform students or amuse them* (or make them feel good about themselves-RT) *or to make them excellent technicians. It is to unsettle their minds, widen their horizons, inflame their intellects, teach them to think straight, if possible."* - Robert M. Hutchins

> *"Education would be much more effective if its purpose was to ensure that by the time they leave school every boy and girl should know how much they do not know, and be imbued with a lifelong desire to know it."* - Sir William Haley

How few teachers understand the importance of convincing their students that they can be whatever they want to be and "turning on" the desire to learn rather than plodding slowly through some canned curriculum.

> *"Knowledge always desires increase; it is like fire, which must first be kindled by some external agent, but which will afterwards propagate itself."* - Dr. Samuel Johnson.

This is the primary function and duty of teachers – to kindle the flame. My children were fortunate in that they eventually ran into a teacher or a professor who really believed in them. He/she was able to throw the switch that convinced them they could do and be whatever they wanted to do, and propelled them into a growing desire for knowledge. Unfortunately, one was a junior in high school and the other two had earned one or more college degrees before that happened. I can only wonder how different the whole educational process would have been for them if this awakening had happened early in grade school. They went to very highly rated schools. I shudder to think about what is happening in many inner city schools today.

"In order to succeed, we must first believe that we can."
- Michael Korda

"Our chief want in life is somebody who will make us do what we can." - Ralph Waldo Emerson

Numerous studies have shown, conclusively, that a teacher's expectations have a profound influence on a student's achievement. I wonder if this doesn't account in part for the high achievement levels of home-schooled kids. The teachers who win awards for outstanding student achievement, like Mr. Gatto, all convinced their students that they could achieve far more than either the students or their previous teachers had expected.

"If we treat an individual as he is, he will stay as he is; but if you treat him as if he were what he ought to be and could be, he will become what he ought to be and could be." - Johann Wolfgang von Goethe

Our educational system is designed to turn out mediocrity. Class progress and lesson materials are geared to the lower third. The rest of the class is bored stiff. The system is designed to turn out cookie-cutter students who have memorized enough to pass a minimum test so schools can keep their pass/graduate ratios up.

"A man must consider what a rich realm he abdicates when he becomes a conformist." - Ralph Waldo Emerson.

I wonder if teachers and administrators ever think about what a ruinous disservice they are doing to the students when they insist on conformity and try to make everyone "equal"? We don't need homogenized children. We don't need a homogenized society. Such a society may be easier to govern and exploit (by the multinational corporations) but it kills creativity and productivity. We need individuals who want to be a cut above and believe in themselves.

> *"The first and most important step toward success is the feeling that we can succeed."* - Nelson Boswell

> *"When we all think alike, no one thinks very much."* - Walter Lippmann.

> *"That which is to be most desired in America is oneness and not sameness. Sameness is the worst thing that could happen to the people of this country. To make all the people the same would lower their quality, but oneness would raise it."* - Stephen S. Wise

> *"To sentence a man of true genius to the drudgery of a school is to put a racehorse in a mill."* - Charles C. Colon

Ask Albert Einstein about that one. It isn't just the genius that is negatively affected by the drudgery of our present school system. It is everyone involved. The system seems designed to turn children off to school and learning. It will remain so as long as the NEA virtually dictates curriculum and the liberals/socialists in the government back them. It is no wonder that people like Churchill and Einstein were bored silly and did poorly in school; our current system would turn anybody off.

> *"Unless the young man looks around for himself and uses his own power of observation and proves the assertion to be the falsity that it is, he falls under the spell of the misguidance and succumbs to a life of drudgery."* - Edward W. Bok

Instead of training students to think, we are training them to parrot the proper answer according to whatever is politically correct at the time. Our latest history books are full of false statements and deliberate omissions. One textbook defined pilgrims as people who took long trips.

> *"Patriotism is easy to understand in the United States. It means looking out for yourself by looking out for your country."* - Calvin Coolidge

Our educational system has trashed patriotism, our history and the moral and spiritual values that made this country great in order to foster a one-world mentality. It has taught our children that the most important thing is to satisfy their own desires and that they can determine what is right and wrong by what they feel is right and wrong for them. It is no wonder our kids are shooting one another if they feel that they have been insulted.

> *"When ancient opinions and rules of life are taken away, the loss cannot possibly be estimated. From that moment we have no compass to govern us, nor can we know distinctly to what port to steer."* - Edmund Burke

In place of proven values and ethics, we now have "if it feels good, do it" being taught. Values clarification classes are designed to throw out the values that parents have taught their children and substitute for them politically correct liberal/humanist ones.

> *"Educate men without religion, and you make them but clever devils."* - Duke of Wellington

WHO SETS THE AGENDA?

> *"A man ought to read just as inclination leads him, for what he reads as a task will do him little good."* - Dr. Samuel Johnson

Obviously we have to give our children some guidance in what they read and we want them to know some basic information and acquire some basic skills. We can't just turn them loose in a library at age six, although many, if not most, would come out better educated than most of our grade-school graduates now. The point in the above quote is that we need to encourage what the child's natural interests are, not try to mass produce identical cookies.

We don't just need a drastic overhaul of the current system. We need to throw out the faulty concepts that the whole system is built upon and start over. Until that can be accomplished, there are some steps we can take to dramatically improve things. The first thing that needs to happen is that competition must be brought into the educational arena. Monopolies just don't work in any area. No matter how well intentioned, they become inefficient, bloated with administrative overhead, and stuck in ruts. Innovation is discouraged, not rewarded. If you want the extreme example, take a look at the Soviet Union, where the government had a monopoly on everything, and nothing worked very well or at all because there was no competition and there were no incentives.

We need a voucher system that parents can use *anywhere*! That means public schools, private schools, religious schools, home schools, and co-ops. This is the only way the public system will ever shape up. The NEA, bureaucrats, teacher unions, and administrators will do everything in their power to prevent the needed changes. The NEA is one of the most powerful lobbies in Washington, but the people are ultimately more powerful if we get our act together. The first step is to get the federal government out of education all together. An obscene amount of money disappears into the federal educational apparatus and accomplishes nothing. Then there is the little factor that the federal government has no constitutional authority or even permission to be involved in education in the first place. That belongs to the states and to the people.

You probably noticed that I included religious schools in the vouchers. Doesn't that violate the separation of church and state? No, it doesn't. In the first place, there is no separation of church and state in the Constitution. This whole quagmire is discussed in the chapter on the Constitution. Briefly, the First Amendment says that Congress shall make no laws regarding the *establishment* of religion. This had everything to do with the government establishing a state religion, not with removing the spiritual basis of morality from government which the founding fathers said was essential to our survival. You have to be a totally muddle-headed lawyer who has been thoroughly indoctrinated in a particularly bizarre view of the Constitution to believe that giving a parent a voucher to use *wherever* they choose amounts to trying to establish a *particular* religion by the government.

Neither Congress or the state legislatures should be afraid of what the courts might do. You never know. If they did throw out the possibility of using vouchers at religious schools, we would just have to elect a president who would appoint judges who could read plain English and understand it. Besides, there is a prevailing precedent for such a use. The GI Bill could be applied to religious schools. What's the difference between that and vouchers? Pell Grants, government loans, and a whole variety of programs are available to students in church-related schools. Why not vouchers? Mostly because the educational bureaucracy knows they can't stand the competition and the kingdom they rule over would have to change drastically and begin educating our children in order to survive. The educational establishment is violently opposed to vouchers for any use and they are using the religious issue as one tactic to fight them. Then there are the politicians who want to see our children continue to be indoctrinated with their particular world view at our expense. They need to be replaced.

The whole idea of a group of kids, all the same age, spending endless hours in hard wooden chairs watching a teacher talk from some dictated course outline is absurd. You couldn't design a worse approach. This approach is incredibly boring, fosters competitiveness, rather than cooperation, stifles any creativity, and is incredibly inefficient. We shouldn't even have compulsory public education in the first place. *What*? You read that right. Before we had compulsory education, the literacy rate among citizens was 98%. What is it now? Don't ask. Back then, parents were responsible for seeing that their children could read and write. They did, with considerably fewer resources than we have now. Radical? Yes. Responsible? Yes. The whole basis for compulsory education is the liberal idea that the government knows better than parents what is best for their children and can do a better job of raising them. The government does a fantastic job on the Indian reservations, right? Our schools, especially the inner-city schools, are models of efficiency and productivity, right? With help (control) like that, who needs enemies? We need to teach our children skills, facts, and methods. They need to be taught abilities that will make them feel good about themselves *because* of their capabilities. Instead, the current system is far more interested in just making them feel good, even if they can't read or write. The results are illustrated in the following two paragraphs.

> *"Patti Johnson, President of the Parents Education Network in suburban Denver, complained that her son never received spelling lists or participated in spelling bees. The reason? One teacher said, 'correcting spelling would stifle his creativity,' while another public school staffer said misspellings 'would hurt his self-esteem'.*
>
> *'They eliminated spelling bees so no child would feel like a loser.' Johnson observed, 'Competition was a dirty word in the classroom. Wait until these students enter the real world and discover competition and losing — then where will their false self esteem get them?'...*
>
> *"One suburban Denver middle school came up with a spelling list of words that students are required to learn how to spell. Sixth-graders were expected to be able to spell a, all, any, go, me, you, he, to, and we. Among the words on the seventh-grade list were eat, end, us, boy, top, car, red, far, and tree. Eighth-graders are expected to spell such difficult words as am, ball, blue, dog, girl, dry, yes, box, and ship." - McAlvany Intellegenct Digest,* September 1994

Is this education? In colonial days, sixth-graders were reading Thomas Paine.

I am not talking about doing away with publicly funded education, just compulsory education of a particular kind. The present one doesn't work. Parents need to have completely free choice regarding the method and content of their children's education. That's the American way, or at least it was. That is what the American Revolution and the Constitution were all about, freedom of choice and a minimum of governmental interference. Somewhere along the way this ideal has been nibbled away into mouse droppings by the inherent drive of politicians and bureaucrats to control everything.

Education or Indoctrination - 121

My wife and I were fortunate in being able to overcome many of the faults and the liberal political propaganda of the educational system with our own children. It is getting harder and harder to do that, with all of the garbage that is now coming from the media to reinforce much of what goes on in the schools. Until some form of competition forces the public schools to begin to actually educate our children again, there are some things you can do. Home school your children if at all possible. In most states you don't need to be a certified teacher to home school your own children. There are many excellent lessons, guides, structured curricula, and other resources available to you. Three hours a of day home schooling will result in higher test scores, more rapid progress and a much more rounded education than six to seven hours in almost any public school. Home schooling will develop your child's creativity and allow you to emphasize the things that he/she is most interested in. The socializing that public school bureaucrats claim isn't provided in home schooling can easily be supplied by cooperative activities and classes with other home schoolers. There are such co-ops in most cities and towns already. Recent tests show that home schooled kids end up five to ten *years* beyond their peers in their ability to *think!* If home schooling is not a possibility, try to get your children into private schools. Yes, it is an added burden. Which is more important to you, providing your kids with a quality education and protecting them from some of the damage from public schools or taking an expensive vacation? You brought them into this world. They didn't ask to be born. It is your responsibility to provide them with the best education and opportunities in life that you can.

It you have to send them to public schools, fight with the bureaucracy in order to send them to the best school in the system. You will probably be told that you can't do that. Insist that you will do that. Stay on their case until they get sick of you and let the kid go where you want. If that doesn't work, take the kid to the school you want and enroll them. If they won't accept them, refuse to send your kid to an inferior school. Claim discrimination. Picket the school with your child. Invite the TV stations. Make the bureaucrats' lives, well, difficult, until they give in. It's your child and your money. You should be able to send your kid to the best school available. What would happen if everybody did that? *Exactly!*

Folks, this is *your country!* Take it back. These are your children, not the government's. Teach them what *you* want them to learn, not socialist, humanist, liberal propaganda.

The pagans and socialists/humanists at the NEA want total control of your children. According to a recent NEA decree:

> *"The Association believes that federal legislation should be enacted to assist in organizing the implementation of fully funded early childhood education programs offered through the public schools. These programs...should include mandatory kindergarten with compulsory attendance."*

In other words, some kids are getting through the public school curriculum without succumbing entirely to the NEA's indoctrination program, so they want to start at an even earlier and more impressionable age. Also, the system has been such a colossal failure that they want the kids to have *more* of it – classic liberal thinking.

As usual, the liberals want more money, more control, and more hours of the same failed approach. Good education isn't a matter of money. New Jersey spends more money per public-school student than any other state, but it ranks 39th on SAT scores. The District of Columbia is number five in spending per pupil but ranks 49th. Utah spends less per pupil than any other state, only one-third of what New Jersey spends, but ranks fourth in SAT scores. It seems to have more to do with whether liberals or conservatives are running the schools and whether the money goes to administrators or is spent on the students. One Indiana school had top scores but spent only $3000 per year. Bill Bennett went to find out how they achieved these scores with so little money. The superintendent said, "We have so little money that we have to focus on the basics."

> *"It is preoccupation with possessions, more than anything else, that prevents men from living freely and nobly."* - Bertrand Russell.

> *"No social system will bring us happiness, health and prosperity unless it is inspired by something greater than materialism."* - Clement R. Attlee

We are trained by the public education system and the constant bombardment of advertising to be voracious consumers and never satisfied. If we succumb, we shall never be happy, much less free or noble. If we believe that we are only animals without hope or future beyond this life, we shall never be any of the above nor ever know real peace.

> *"Whatever disunites man from God disunites man from man."* - Edmund Burke

Our public education system has put a great deal of effort into disconnecting man from God. Is it any wonder that our schools and society are dissolving in mindless violence?

> *"The higher men climb the longer their working day, and any young man with a streak of idleness in him may better make up his mind at the beginning that mediocrity will be his lot. Without immense, sustained effort, he will not climb high. And even though fortune or chance were to lift him high, he would not stay there. For to keep at the top is harder almost than to get there. There are no office hours for leaders."* - James Cardinal Gibbons

Why are our public schools not teaching our young people such facts of life instead of teaching them that they are victims of something or someone, which will never get them anywhere? If you have to send you children to public schools, try to supplement their education, if that's what you can call it, at home or elsewhere, as much as possible. While the school setting is trying to destroy your childrens' self esteem and expectations, try to instill in them some of the following ideals:

> *"The quality of a person's life is in direct proportion to their commitment to excellence, regardless of their chosen field of endeavor."* - Vince Lombardi

"Happiness lies in the joy of achievement and the thrill of creative effort." - Franklin D. Roosevelt

"Accept the challenges, so that you may feel the exhilaration of victory." - Gen. George S. Patton.

"Chance favors the prepared mind." - Louis Pasteur

"The difference between the impossible and the possible lies in a man's determination." - Tommy Lasorda

"Do not let what you cannot do interfere with what you can do." - John Wooden

"The harder you work the luckier you get." - Gary Player

"The spirit, the will to win, and the will to excel are the things that endure. These qualities are so much more important than the events that occur." - Vince Lombardi

"Paralyze resistance with persistence." - Woody Hayes

"The only limit to our realization of tomorrow will be our doubts of today." - Franklin D. Roosevelt

"What the mind of man can conceive and believe, the mind of man can achieve." - Napoleon Hill

"The only thing that stands between a man and what he wants from life is often merely the will to try it and the faith to believe that it is possible." - Richard M. Devos

"Whether you think you can, or think you can't – you are right." - Henry Ford

Education or Indoctrination

"Far better it is to dare mighty things, to win glorious triumphs, even though checkered by failure, than to take rank with those poor spirits who neither enjoy much nor suffer much, because they live in the gray twilight that knows not victory nor defeat." - Theodore Roosevelt

I have included quite a few similar statements. I suggest that you have your children memorize these one at a time, perhaps one a week and give them a reward when they do. Have them look up the people who made the statements in the encyclopedia and tell you about them. If you can ingrain these concepts in your kids' minds early in life, there is little that can stop them from succeeding later on.

I also suggest that you read to your children as much as possible, beginning in the womb. After they are born, read to them daily while feeding or holding them. Make their earliest experiences with reading part of the pleasant things that they desire the most. Make that connection and they will have a strong desire to read for life. Keep reading to them until they won't let you anymore. Then, try to get them to read to you. Include some of the classics and books on freedom like those of Thomas Paine. Take time to discuss the ideas and concepts in the books. Let them see both parents taking time to read good books. Kids do what their parents do, more than what their parents tell them.

If you are interested in what can be done to fix the public education system and make it work, read Wayne Green's book, *We The People Declare War On Our Lousy Government*. Despite the iconoclastic title, it is mostly about ways to get the public schools to actually produce usable products that work. Frankly, I don't think the system can be fixed unless politicians can be elected who will honor the oath they take to defend and protect the Constitution and return full control of education to the local school boards and the parents. The teachers will have to form a rival association to the NEA that is actually interested in real education for the children in order to get control of the curricula and be able to teach what they want to, the way they want to.

Even then, the whole approach to education in the sterile, boring, isolated classroom settings we have now remains the worst method available. As long as you are stuck with it, if you want to return control over your schools to your community, elect school board members who will put in real curricula and fight the federal intrusion and the NEA's indoctrination program. Insist that the money be spent on the children, not administrators. Demand the latest in technology to make learning fun and useful.

If you would like some information on an excellent (I think it is the best) computer based home or private school curricula, write to the publisher of this book or use the order blank in the back for free information.

Chapter Five
Crime and Circumstances

"Whom the gods would destroy, they first subsidize." - George Roche

WELFARE

"The only four-letter word that still shocks people these days is WORK, plain old everyday work." - Thomas Sowell, Economist

If you are wondering what welfare has to do with crime, read on. That we have a real mess with the welfare system is well known. That it is also a dismal failure in most respects is fairly well accepted. Few people, however, realize just how destructive the welfare system has been to everyone involved, including you and me. Admittedly, it has helped some people get back on their feet. These people are a small minority of welfare recipients. For the vast majority of people on welfare of any kind, the system has been more destructive than helpful. There is also the fact that federal welfare is outright unconstitutional. Nowhere in the Constitution is the federal government given the power to confiscate resources from a farmer in Idaho and give them to someone in Chicago to pay their rent. The taxing powers were there only to pay the necessary expenses of operating the federal government within the strict limitations placed on it by the Constitution. This will be discussed in the chapter on the Constitution.

I could cite case after case that would aggravate you a great deal. Many women on welfare freely admit that they had some of their children specifically to collect more welfare. These children were often fathered by different men for each child. Frequently, the women lose custody of the children because of being in jail for various offenses, sexual abuse of the children by boyfriends or among the children themselves, or serious neglect. Then, they use "free" legal services paid for by your tax dollars to try to regain custody so they can continue to collect the AFDC payments.

Some women have been given house after house, often very nice homes in the scattered housing programs, only to be forced to move becuase the homes were trashed and damaged so bad as to be unlivable. Others were forced to move because of repeated police calls to the homes for violence or illegal activities. Many of these women freely admit that they could get jobs, but they can't get as much money or benefits by working. Some even say that welfare should be eliminated, but they are going to keep on getting a free ride as long as they can. Some blame the government for being stuck on welfare and others blame their mothers for setting the example of living on welfare.

It would be bad enough if your tax dollars were only going to the recipients, but only twenty percent of the money gets through to them. Eighty percent of the money you pay to support the welfare system goes to the huge bureaucracy that has grown up around these people.

To say that welfare is an enabler is to understate the problem. It is a powerful incentive to get on to and stay on taxpayer subsidies. It provides strong incentives for women to have children and raise them by themselves, thus perpetuating and expanding the system. Besides the direct cost of raising these unwanted, or at least "enabled" children, there are also costs of continuing adult welfare, prisons, crime, and other costs to society that these children incur as they mature. It has got to stop. Not only are the people on welfare being callously used and abused by the politicians, the U.S. economy simply cannot afford it. Just how "compassionate" is it for the child who is conceived to collect money for welfare, grow up mostly on the streets with no father and a mother on drugs, without hope of a decent education, job or future, join a gang for survival, and likely spend a good part of his life in prison? Some compassion. "That's not the way it is supposed to work", complain the liberal apologists. They have nice sounding theories; welcome to the real world. Trying to tinker with the current welfare system just won't work. It's foundation is based on false assumptions. It has to be scrapped.

> *"Real Poverty is less a state of income than a state of mind."* - George Gilder

Lyndon Johnson started the War on Poverty and the war in Vietnam, although Kennedy had sent in advisors. We have lost both wars. Since 1965, *you and I* have paid out more than *five trillion* dollars in welfare benefits out of our pockets, and there are more people living in poverty than before. Because of the incentives built into the programs, we have actually caused more poverty. What ever the government subsidizes, it gets more of. To put this amount in perspective, it is more than the cost of defeating both Germany and Japan in World War II. It is also more than our current national debt. Despite the dismal failure of the welfare fiasco, the Democrats want to cut the defense budget and increase welfare.

> *"Pennies do not come from Heaven. They have to be earned here on earth."* - Margeret Thatcher

Here is another choice example of the welfare bureaucracy lunacy:

> *"Recently a 50-room building in a blighted section of Spokane was completely renovated, repainted, carpeted, and had new furniture and clean bathrooms. Many new residents had federal rent subsidies.*
>
> *"To keep the apartments safe, rules were strictly enforced: no alcohol nor drugs, residents had to perform chores, obey curfews, and submit to drug testing. It was a pleasant place to live. But then HUD (Housing and Urban Development) decided that the strict rules discriminated against DRUG ADDICTS and alcoholics, so they stopped subsidizing the complex, leaving management no choice but to abandon the rules in order to get the subsidy.*
>
> *"Within weeks, several rooms were being used for prostitution and drugs, human waste could be found on the floors, and the building went downhill."* - Howard Ruff, *The New Ruff Times.*

It is time this kind of nonsense came to a screeching halt. This is your tax money at work. The government is subsidizing drug addicts and prostitutes. Whatever the government subsidizes, it gets more of. The owners of the building were denied the right to control their own property if they wanted to survive financially. The result was a destruction of the property and financial distress to the owners anyway. Most of the people there lost the right to live in a decent place all because HUD *thought* that the policies necessary to keep the place livable *might* discriminate against drug addicts. Does welfare increase crime? This is another example of nanny government at its best.

> *"Whenever government decides to solve something, we have learned to be wary. The cure may not always be worse than the disease, but it usually bigger and it costs more."* - Ronald Reagan

After a considerable amount of research, I have come to the following conclusion: either the liberal politicians who created this mess are hopelessly naive about human nature and economics, or they have deliberately created huge breeding farms (constituencies) whose inhabitants have to continue to vote for them in order to keep the subsidies coming and thus keep the politicians in office. Either way, these types of politicians have no business handling our tax money. Either they are incompetent or they are buying votes with your money. What else is new? They buy votes with our money through pork barrel projects, free mailing, and dozens of other means. However, these don't cause anywhere near the amount of human misery that the welfare system does.

The federal government has no constitutional authority to confiscate and redistribute assets that citizens have lawfully earned. Redistribution has nothing to do with raising the needed funds to pay for the necessary expenses of legitimate (constitutional) government expenses. Trying to implement this idea of redistribution would have had our founding fathers reaching for their muskets. Forget block grants to the states; the money should have never left the states in the first place. The federal government should get out of the welfare business entirely, even if welfare wasn't unconstitutional. Congress has proven that their "fixes" usually make things much worse for everybody. The states and private charities can do what needs to be done far cheaper and more efficiently. We should stop all federal welfare and reduce federal taxes by that amount, so that states can raise taxes according to their individual needs and their citizens' wishes. Let different states try different approaches, as is their right under the Tenth Amendment, and see what works best. This is part of the genius of how the Constitution was written and violating this separation of powers has been incredibly costly, in more than just money. Back to the Constitution!

CAUSES OF CRIME

Swift apprehension and punishment for criminal acts, including long term removal, are necessary to protect society and provide some amount of deterrent. However, the most effective approach to preventing crime is to deal with its root causes.

Let's take a look at some of the changes that have affected crime rates. The Census Bureau reports that births by unwed mothers rose more than seventy percent between 1983 and 1993. A lot has been written and said about the disastrous effects of fatherless homes. Most of it is accurate or even understated. The effects of the absence of a good male role model have been discussed quite thoroughly. What hasn't been discussed very much is the importance of bonding and the search for acceptance by a male role model by children in their younger years. This has led children to find male role models and acceptance in gangs.

Less than thirty percent of Black children, forty percent of Hispanic children and sixty percent of White children live in traditional families. Fifteen percent of all children live with a step-parent or step-sibling. As a result of the lack of traditional family structure among such a high proportion of Black families, they represent a highly disproportional segment of the prison population. This broken family structure is not a racial characteristic, as shown by the tight family and tribal structures in Africa. It is a result of the government's strong incentives for having one-parent families, especially single teen-aged mothers. While discrimination undoubtedly plays a part in the high prison percentage, the lack of good male role models and male discipline in the early years appears to make the most difference, regardless of race.

> *"Seventy-five-per-cent of all mothers with children between 6 and 17 are now in the work force."* - Business Week, January 2, 1995

The vast majority of people in prison come from single parent families, nearly all of them with an absent adult male. While families with two working parents have a much smaller percentage of such problems, the percentage is still higher than for the more traditional one-parent-at-home family. Contrary to the liberals' claim that poverty causes crime, it is the government's policies that encourage teen-age and single mothers to get pregnant to collect welfare who usually end up living in poverty. These policies help cause *both* the poverty and the crime. Unfortunately, these young single parents usually have virtually no parenting skills and little education.

The current crime and drug problems, that are most prominent in the inner cities, are a direct result of the social schemes of the liberal politicians and the attacks on traditional families by the media and the radical feminists.

Until the government gets out of the business of providing incentives for single women to get pregnant, financially penalizing two parent families, providing very inferior and largely useless education, and encouraging dependence on government handouts, all the prisons we could possibly build at incredible expense will do little to affect crime. The attacks on religion by the government, particularly by the courts, are also suppressing the spiritual values that are the foundation for any society's moral values and actions.

DRUGS

In addition to the fact that the federal government has no constitutional authority to provide individual health care to anyone except its own employees or former employees (like veterans), it continues to try to run drug rehabilitation programs that don't work.

A few years ago the government's major program in Kentucky had a success(?) rate of about five percent, at a tremendous cost per person treated. Contrast that with the Teen Challenge program founded by David Wilkerson, with a success rate of over eighty percent. Private enterprise strikes again.

Out of each dollar allocated for treatment in the federal programs, most goes to administration and overhead as usual. What little ends up actually being spent on the individual with the problem does virtually no good anyway. Why don't the state and federal governments simply send people to Teen Challenge and similar programs? Well, first of all, all the people who work in the various government programs would be out of a job, and we can't have that, even if they are wasting a tremendous amount of resources. Then there is that little church-state thing. We can't have the government paying any Christian-affiliated group money, regardless of how successful they are at solving the problems. Right? Wrong! The state and federal governments pay Catholic, Baptist, and other church-related hospitals all the time to care for patients through Medicare, Medicaid and other programs. What about that "wall" of separation of the church and state? Well, there is no such thing in the Constitution, but that subject is dealt with in Chapter Twelve.

It does absolutely no good to fill our overcrowded prisons with people who have been arrested for possessing a small amount of an illegal drug. We need that space for hardened and dangerous criminals. For far less money, the state could put these people in programs like Teen Challenge. The failure rate for drug programs in prison is almost total. These people get out and return right back to a life of crime in order to feed their drug habit. This approach is nonsense. It is ridiculously expensive and accomplishes nothing. For what it costs to lock up one addict, they could send ten through a program like Teen Challenge, and eight of those ten would be drug free from then on. The government could pay for the treatment just like they do in a church-related hospital. There is absolutely nothing in the Constitution that could possibly be construed as forbidding the payment for such care, unless the judge is living in Alice's wonderland. It is quite simple, really. Spend lots of money for virtually nothing, or spend far less money for effective treatment. Are there any politicians with common sense out there? If you insist on confinement rather than treatment, see the section below on suggestions for improvements.

States, not the federal government, have always had the Constitutional right to provide for the health and welfare of their citizens, at least until the Roe v. Wade decision was handed down. That decision was a major blow to this right, and Hillary's health care commune was a blatant attempt to take our few remaining rights away from us. I can only hope that at least a few states will adopt the approach outlined above. The ACLU or some other group will challenge such action and shop around until they find some liberal federal judge who will say such action violates the separation of church and state. I think that the states can win by demonstrating that government payments to hospitals are a clear precedent. However, if the federal courts should decide otherwise, I suggest that the governor and the legislators ignore the courts and go ahead anyway, citing the Tenth Amendment, which is much clearer and more directly to the issue than the smoke screen about the phony "wall". Is a judge going to try to throw the governor and the legislators in jail for contempt? He and what army? He could use federal marshals, and then the governor could call out the national guard. The federal government could nationalize the guard, and then what? Hopefully, the people of the state would be so up in arms that Congress would have to act. As Abe Lincoln said, a little rebellion now and then is a good thing. In addition to something like the Teen Challenge program, states should also be free to experiment with other approaches to the drug problem, such as carefully controlled maintenance programs like those being used effectively in other countries. With various states trying different things, we have a far better chance of coming up with programs that work. What we have now is a horrendously expensive disaster.

The above scenario is, of course, an extreme example. There is no need for a confrontation to go that far. Maybe common sense would prevail. However, I wanted to make a point. It is time for the states to tell the federal government to get lost and start solving their own problems. Federal programs seldom work for almost anything you want to mention. Moreover, most of them are blatantly unconstitutional under the Tenth Amendment. We need to start supporting what does work. We can no longer afford to do otherwise.

If the governors and legislators of the various states don't stop looking to Washington to solve their problems, then we need to elect different people to our state governments who will solve the problems themselves and stop whining about not getting enough federal money from Washington. We need to keep that money at home in the first place.

OTHER CAUSES OF CRIME

There are many other causes of crimes, including greed, mental illness, breakdown of moral values, encouragement of violence by the media, and advertising, to mention a few of them One particular area that I would like to mention is the bonding/nurturing problem. We now know that bonding with parents, particularly with the mother, is very important to normal mental and social development. The first twenty minutes after birth are especially important. Unfortunately, the baby is often removed during this critical time for some reason.

We are astounded and dismayed when we see fourteen year old kids, with a total lack of emotion or remorse, brutally kill people for the slightest irritation or for no reason at all. Some new light is being shed on this through research. Apparently, many of these kids had little or no bonding with either parent and had constant stress in their earliest years because of a mother on drugs or alcohol, child abuse, or simply because their most basic needs were neglected. This stress causes nearly continual adrenal gland stimulation which, in turn, causes physical brain damage. This damage appears to affect emotional responses and self-control the most. These kids do not have the capacity to respond emotionally to other people, so they kill without any feeling or conscience. Apparently the damage cannot be repaired.

This is not to excuse such behavior. These kids are still responsible for their behavior and must be removed from society. However, maybe this problem underlines the seriousness of the need of proper family support and care for children that our current system is destroying. In case you are wondering, this love, support, bonding, and care cannot be provided by child care centers or even substitute parents. It must come from the family.

SUGGESTIONS FOR IMPROVEMENTS

We simply cannot keep on building more and more prisons. We don't have the money for one thing and it isn't necessary, for another. Prisons should be reserved for those who are dangerous or unable to control themselves. We have the technology to electronically confine people to their homes, if they present no violent threat to society. Why should you and I have to work to feed them, clothe them, house them and provide medical care? The expense is immense. Confine them to their homes. Place whatever restrictions upon them that are appropriate. Let them pay their own way. Let them figure out how to support themselves. If they break the rules, chain them to posts in their houses. It's their choice.

We pay too much attention to the criminals and not nearly enough to the victims. Every crime against property or people should require appropriate payment to the victims by the perpetrators. Simply sending someone to prison for a given time is not much of a deterrent anymore, if it ever was. We should require criminals to pay for the damage caused by their crimes even if it takes their whole lives. They should never be free of the system and some control over their conduct until the damages are repaid. If that takes them their whole life, tough.

Some commercial prison designs have open structures that significantly reduce the cost, the crime, the sexual and physical assaults, and other problems. This is another area that private enterprise can handle better and more cheaply than the government can.

We need to revise the criminal justice system. One of the reasons that there are a disproportionate number of poor people in prison is that the outcome of a trial often depends more on the skill/cost of the lawyers than on the facts. In most European countries, trials are much quicker and cheaper. Both the prosecution and the defense are required to do whatever they can to find the truth and achieve the most appropriate result. None of these court systems would tolerate an OJ trial. Justice is quicker and more sure. Plea bargaining is virtually unknown. If you do the crime, you do the time.

Most important of all, we need to rebuild the family structures and the moral values we once had, which can only be built on spiritual values. If you don't think you need a spiritual base, try to build a moral structure without any commonly accepted authoritative standard of values. Russia and many other "former" communist countries have had enough sense to ask for all the Christian educational material and help they can get. After telling people for seventy years that they are the result of an unlikely series of biological accidents, that they have no purpose or meaning except to further the leaders' agenda, and no real meaning, these countries have no moral values. Without moral values, a society will have rampaging crime and chaos. Conducting business is almost impossible. Contract enforcement is almost impossible. If you want a picture of what we will look like in the not too distant future, unless changes are made, take a good look at Russia today.

RACIAL DISCRIMINATION

This section could have gone into several chapters, but since the liberals and many "spokesmen" for minority groups blame the high crime rates among minorities, and blacks in particular, on poverty and discrimination, I chose to put it here.

Black teen-aged males have a much higher unemployment rate than white males of the same age, with other minorities usually falling somewhere in between. Why? Is it discrimination? Yes, to a certain extent, but discrimination based on what? Race? Sometimes. However, a lot of it is based on education and skill levels, which are often inferior in inner city schools. Quite frequently, the lack of employment is based on communication or language skills. Any job that requires public contact usually requires clear communications. Fortunately, some educators and others (mostly black) are starting to recognize just how important it is in the job market to be able to communicate clearly and effectively and are taking steps to help people do so. Talking in a hard-to-understand slang just doesn't work. Like it or not, business in most of the world is white-dominated. But regardless of who runs what, effective communication skills and abilities are essential to any degree of success in almost any area. Even Michael Jordan wouldn't get all those product endorsement opportunities if he couldn't speak clearly and effectively.

Another area frequently overlooked, often deliberately, is the family structure and the culture. Many Hispanic children live in the same neighborhoods and go to the same schools as Blacks. However, they are more likely to have two parent homes. The crime statistics reflect the difference. Many Asian families may be just as poor but far fewer are on welfare, percentage-wise, and family identity and structure are much more important. Discipline is usually much stronger. Hard work and achievement are strongly emphasized. Excuses for failure because of discrimination, which is very real, are not acceptable. Crime statistics are much lower. Many of the Vietnamese refugees came here with only the clothes on their backs but virtually none are on welfare and today a high percentage are prosperous business owners and professionals even though they couldn't even speak English when they arrived. They refused to let the government make them a dependent group. The overseas Chinese have been fiercely discriminated against in many countries for thousands of years, yet they often end up being the achievers and owners in many of the societies. Racial, ethnic, and other kinds of discrimination are real and they harm everyone. Ask the Bosnians. But if anyone uses discrimination as an excuse to keep from striving to be all they can be, they are the big losers. I am not using these comparisons to put down any particular group, but to emphasize the extreme importance of family structure and expectations.

Children raised by single parents-too-soon often lack the self-discipline, focus, and commitment it takes to hold a job, much less succeed at it. You have to show up for a job consistently and on time if you want to hang on to it. It is a real indictment of our educational system that McDonald's and other fast food places have done a lot more to teach the necessary skills and prepare young people to enter the job market than our incredibly expensive public school system has. Yet, the government keeps raising the minimum wage and pricing marginal employees out of the job market. Then where do these entry level people get their start and training?

> *"It is old-fashioned, even reactionary to remind people that free enterprise has done more to reduce poverty than all the government programs dreamed up by democracy."*
> - Ronald Reagan

Affirmative action was supposed to help solve the job discrimination problem. Has it? By and large, no. It has helped a surprisingly small percentage get jobs or promotions that they might not have otherwise obtained. Sometimes this has led to frustration, stress and other problems for people promoted to positions they were not qualified for. It has also been a burden on businesses who hired underachievers who were largely dead weight. Many businesses have gone out of business because of affirmative action or unjustified job discrimination complaints with a loss of jobs for everyone, resulting in a loss to the whole economy, including tax revenues. You can't compete against very industrious workers in Asia with people who don't produce.

Should there be laws against job discrimination? To a certain extent, yes. One of the positive effects of all of these laws has been that more people of different races have been working together and getting to know one another better as individuals, which has helped race relations to some extent. However, reverse discrimination has only worsened race relations as anyone with any common sense would deduce. As long as the government emphasizes race to any significant degree, the longer it will take to achieve any significant racial harmony.

Martin Luther King said, "I look for the day when men will be judged by their character, not by their race" Our whole society would be far better off if this were the case. It never will be the case in this world completely, but the closer we can come to it the better off we will all be. How many Washington Carvers never had a chance to contribute to education because of racial discrimination? How many Willie Mays and Jackie Robinsons never got the chance to compete in the big leagues because of discrimination? How many doctors and scientists .... well, you get the idea. Those that were denied their rightful destiny lost the most, but we all lost to some extent.

> *"Whoever seeks to set one race against another seeks to enslave all races."* - Franklin D. Roosevelt

Until the government is completely racially neutral, and laws are completely colorblind, race will always be a significant problem in this country. Whether it is "normal" discrimination or reverse discrimination, any discrimination produces resentment, hostility, and hatred.

As Bill Bennett, along with a lot of younger black leaders and spokesmen have said, "Quotas exacerbate racial problems." "Equality means equality." Massachusetts has the most liberal laws and regulations concerning race and they have the most racial tension. As long as we are Blacks, Whites, Asians, whatever, we can never be just Americans. It should not be "here's a scholarship for an Asian individual," "here's a job for an African American," "here's a grant for an Irish American." The American way is to maximize the opportunities of the individual (freedom), not deal with groups (socialism).

The whole discrimination problem is an emotional one, not a logical one. The differences are pretty superficial. Mostly, they amount to skin color, hair characteristics, facial features and culture. As far as the genetic code is concerned, there are often more differences in the code between two white Englishmen than an Englishman and a Black African. As far as the geneticists are concerned, race is a minor and trivial differentiation.

What makes a person a member of one race or another? If the Englishman marries a black African and produces a child, is it Black or White? Logically and genetically, it is half White and half Black. What if this child marries someone from South America who is half Japanese and half Native American? Their child is one-fourth White, one-fourth native African, one-fourth native American, and one-fourth Asian. What race is he? This is not just an academic question. More and more people in this country are having trouble figuring out what to put down on all of those government forms. "Other" is becoming one of the largest categories. The real question is, what difference does it really make? Why not eliminate racial classifications on those forms and in a lot of other places? As long as the government emphasizes race so much, there can never be racial harmony.

America is no longer a binary (black/white) society. We are becoming a much more polyglot nation. The vast majority of us are mongrels to some extent. I am "half" English plus a lot of other unknown things and "half" German plus a lot of other unknown things. How about you? Are you racially/ethnically pure? I am rumored to have had a Native American ancestor somewhere back there, but we can't find her and it probably isn't true. I would be glad if it were true. Very few "Blacks" in this country don't have some white ancestors somewhere, due largely to slavery and the eventual ban against importing slaves. If the politicians would stop promoting class and racial warfare in this country to keep themselves in office, we could get down to the business of building America and trying to provide real education and opportunity for everyone. Much of the opposition to forced bussing and the push for school vouchers has come from the Black community, but the liberal politicians would rather stand behind their discredited ideology than do something useful and helpful.

The government's scattered site housing program has caused a great deal of conflict. It is not a good idea to concentrate into one area people whose dependency on the government tends to develop a certain irresponsibility and hopelessness. But scattering them has seriously worsened race relations and generated a lot of anger against the welfare system. Solution: scrap the whole system.

If you had scrimped, saved, and lived in a tiny apartment for years until you could afford a house in a nice community, would you resent it if the government gave the house next door to you to someone who lived off of your tax dollars? It really doesn't matter what race that person is, does it? It happens that most of them have been Black, but I really don't think that this is the main point. Like the woman mentioned earlier who had trashed two expensive homes provided to her, the reason the neighbors got up in arms was not her color, but the seventy-two police calls to her house and all the other problems.

People with similar values, incomes, and cultures tend to clump together. This is true all over the world. Trying to forcibly disrupt this basic tenet of human nature can only result in magnifying already existing problems. We are supposed to have freedom of association in this country. Many Black parents would prefer that their kids go to a mostly Black school or the closest neighborhood school. Why should they be denied that right? Why should their child be a guinea pig in the liberals' social experiments? All parents should have a completely free choice of where and what kind of school their children attend. Solution: vouchers.

As I said before, this section could have gone into several chapters. It ended up here because of the very highly disproportionate rate of crime in predominantly Black communities, particularly in the inner cities. This problem was largely created by the government, particularly Johnson's Great Society, at a cost of trillions of dollars. It is a Black problem simply because most of the people killed, injured and affected in the inner cities are Blacks who are victimized by other Blacks. It is a problem for all of us because we bear much of the cost and some of the blame. Many of us voted for the bozos who created a lot of the circumstances. No government program of any kind is going to solve these problems, certainly not midnight basketball games. What a joke!

I am not a sociologist or a psychologist, just an analytically trained engineer and a bit of an amateur history student. In my opinion, many of these problems are not going to be solved until Black (or Hispanic, or White or ...) males take back their families from the government, assume the responsibility for their families, and tell the government to get lost. Will that be easy? No. With the poor education people have received, poor job skills, few if any role models, and all the rest of it, it will be very difficult. Many people will end up suffering. But the present system isn't working. It created the mess. It has to be stopped. I think it would help if leaders, spokesmen and politicians from these various groups started emphasizing family responsibility, community responsibility, and personal responsibility, and stopped telling people they were helpless, worthless victims who need to be cared for by someone else. Apparently, many prominent younger Blacks agree and I have been encouraged by their boldness in saying so.

As I said much earlier, discrimination is the cancer of human history. Racial, ethnic and religious persecution and discrimination have been responsible for hundreds of millions of deaths and unimaginable suffering. For what? What good has it accomplished?

Chapter Six

The Sky is Falling

"The extinction of the human species may not only be inevitable, but a good thing...This is not to say that the rise of human civilization is insignificant, but there is no way of showing that it will be much help to the world in the long run" - The Economist.

"Bad laws are the worse form of tyranny." - Edmund Burke

As a lover of the outdoors – forests, hills, lakes, streams, wilderness, and wild things – I appreciate what has been accomplished in many areas by cleaning up the environment. Lake Erie was a cesspool. It now offers the best fishing in the Great Lakes. The water is becoming so clear, thanks in part to the unwanted help of the Zebra Mussel, that the low end of the food chain for the fish is threatened.

A great deal of good has been accomplished and there is still more to be done. Unfortunately, in true bureaucratic style, the EPA, OSHA and other agencies have gotten the bit in their teeth and are running away from any form of common sense. In California, a farmer faces a year in prison and $300,000 in fines for accidentally killing five kangaroo rats while tilling his own land. To put that in perspective, people can be punished with anything from execution to not even being arrested for killing a person. Accidentally killing someone often doesn't even result in arrest or a civil suit. Apparently, according to the EPA, rats are far more valuable than people. The ability of these agencies to levy fines against people without a jury trial is a violation of due process and the right to a trial by jury guaranteed by the Constitution. The whole point of a grand jury indictment and petit jury trial is to inject some common sense by common people into the process. When this process is bypassed, we have this kind of nonsense mentioned above.

Even sillier is the felonious feather caper that has been in the news lately. A lady picked up an eagle feather outside an eagle's cage at a zoo. The eagle was obviously done with it. She made a little craft project with it and sent it to Mrs. Clinton. The result was a nightmare for her and her family, costing several thousands of dollars in fines, not to mention legal expenses. All this over a discarded eagle feather. The purpose of the law is to protect eagles from being hunted for their feathers, which is a good idea. This kind of persecution of people just to make a point is nonsense, however. There was no question that the feather was discarded. The eagle didn't even know it was missing and, if he did, couldn't have cared less if it was picked up, swept up or blown away by the wind. No one disputes the fact that she did no harm to anything. If she had picked up the feather to throw it in the trash, she still would have been in violation of the law. Actually, Mrs Clinton was the one in clearer violation of the law for possessing the feather but of course, the EPA did nothing about her. The Clintons seem to have a magical immunity from the law. It helps that if, while you are the Governor that the coroner, state police, and state investigators all work for you. It also helps if, while you are the President, that the Justice Department and the FBI work for you. See the book or the video, *The Clinton Chronicles,* for information on how this works.

Then, there is the wetlands fiasco. The number of ducks are increasing due to the preservation of and improvement in wetlands nesting habitat, among other reasons. That's great! However, most of this improvement comes from conservation groups and sportsmen such as Ducks Unlimited. The contribution from the wetlands laws is highly questionable. The implementation of the law has been absurd in many cases. Farmers are being fined for plowing depressions in their fields that may hold water for a few days if there are heavy enough rains. You can't hatch duck eggs in three days even if the time of year is correct. Ducks have more sense than to build nests in such a place even if the bureaucrats don't. Not even a frog would live near these depressions waiting for some water to show up. Neither can the farmer tile and drain the depression so that his crops won't drown. The government would really crucify him for that. In other words, private property rights guaranteed by the Constitution don't really exist anymore. I am all for preserving *real* wetlands that provide habitat for wildlife. If wetlands are that important to have, let the government buy them. Let everyone pay for the land if it benefits everyone, not the one landowner who was unlucky enough to have a pothole on his farm.

There was a case where someone bought some commercial property to build a store on. He hauled in some dirt to level the back of the property. Sorry Charlie; there was a little ditch along the back edge of the property that could overflow onto the property during heavy rains. The water usually receded shortly after the rain stopped. That property had about as many attributes of wetlands as a street in New York City that floods when it rains hard enough. But, no matter. Water flowed onto the property occasionally under unusual circumstances. The owner was fined and made to remove all the dirt he had brought in. The value of the property plummeted. The moral of the story is, don't buy any property unless it is above the five-hundred-year flood level and is already well drained.

Under the Clean Water Act, a man was sent to prison for building some duck ponds to provide more breeding areas as part of a wildlife sanctuary. Isn't that one of the main purposes of the act? A couple in New York was fined $30,000 for building a deck on their home that cast a shadow on a "wetland". In case you didn't know, a wetland is anything that is dry less than ninety-six percent of the time. If your lawn is wet five percent of the time, and whose isn't, you own a wetland. Be careful what you do with it, or you might end up in jail. If you think that there is no possibility of being fined and/or going to jail, so did thousands of people who have been persecuted by the bureaucrats. If you think that I am kidding about being careful what you plant, there is the case of a woman in Wyoming who was barred from planting roses in her own garden.

Somehow some common sense needs to be injected into the runaway bureaucracies' decision-making processes. All too often, these bureaucrats sound like Orwell's classic example of "pigs is pigs is pigs". First of all, the Constitution gives Congress, not bureaucrats in the executive branch, the authority and responsibility to make laws. As usual, Congress has abrogated its responsibility by passing sweeping generalities and letting the bureaucrats come up with the regulations. That way, when people don't like the results,

Congress can blame the agencies for being overzealous and claim that they really didn't intend things to turn out that way. Somehow we have to get back to the original idea that Congress passes the laws and the executive agencies simply administer them. We also have to return to the basic principle that underlies all of our freedoms – a trial by jury – for any infraction. When any government agency becomes legislature, law enforcement, prosecutor, judge and jury all in one, Big Brother has already arrived. Ask Randy Weaver, or the survivors of Waco, or some of the FDA's victims about how that works.

The media, the environmental extremists, and the government seem to have a ready supply of "scientists" and "experts" to come up with all kinds of scare stories which the press usually blows all out of proportion to make "news" out of them. I have bounced around a large university for too long to believe these people or most of the media "news" they manufacture. There are usually two kinds of scientists: those who do and those who talk. Scientists who get published in respected journals must have their work scrutinized by peer review committees, and the process is tough. In university circles, peer respect is unbelievably important. Those who can't get published because of poor research or ability seem to end up as witnesses in front of congressional committees and "experts" for various groups. These are the people the media like to seek out, because they normally sensationalize things to get attention. Serious scientists usually avoid this kind of exposure like the plague. The Alar scare was a media circus, but it was phony as a three-dollar bill and a lot of apple growers went bankrupt because of it.

PCB

There is no solid scientific evidence that PCB's cause cancer or any other disease in anything like the concentrations that people might be exposed to. People who worked with the stuff for years, got it all over them, breathed it, and probably ingested more of it than you or I would over many lifetimes, suffered no known harm. Yet, we have spent billions of dollars cleaning up the stuff. I would rather be safe than sorry, like most anyone else, but it is long past time that we demanded some actual scientific evidence before launching these massive assaults on problems that might not even really exist. The whole PCB panic started when some people in Japan got sick after eating some rice that was heavily contaminated with PCB. This was a unique accident. As a result, the EPA banned PCBs, which are extremely valuable in many applications, particularly electrical equipment, because they are very stable and *non-flammable*. But, you ask, should we take any risk, no matter how small? That depends. How many people will be killed or injured because the substitutes do not have the same safe properties that PCBs have? Even that begs the question. It turns out that the people in Japan did not get sick from the PCBs in the first place. About half of the contaminant was not the PCBs but other highly toxic chemicals that had been mixed with the PCBs. The sickness was proven to have come from these other chemicals, not the PCBs.

Did a red-faced EPA reverse its ban and admit their mistake? Of course not. So, we have to pay a lot more for substitutes that are a lot less safe. No bureaucracy is going to admit a mistake if they can possibly avoid it, no matter what it costs the taxpayers. The bureaucrats certainly aren't going to do anything that might reduce the size of their empire. Most people aren't aware of the fact that the more regulations an agency can write, the more people it can employ to enforce and deal with these regulations. In a government bureaucracy, your pay grade depends largely on how many people you have working under you. The more regulations you can get on the books, the more people you can hire and the higher your pay will be. That's one reason why bureaucrats generate truckloads of nit-picking and largely useless or counter-productive regulations every year.

Just in case you're wondering, PCBs are not the environmentally long-lasting stuff the extremists make them out to be. There are several forms of bacteria that can and do break them down into simple harmless substances in a few days.

DIOXIN

According to the sensationalized journalism, dioxin is the most toxic chemical known to man. There are only a few dozen things wrong with that statement. First of *all*, dioxin is not a chemical, but a class of about seventy-five different chemicals with very different properties. Next, there isn't one bit of scientific data demonstrating that dioxin(s) causes cancer, birth defects, or miscarriages in humans. The only disease that is known to result from very heavy exposure to dioxin is a skin rash, which is easily treated. Several groups of people have been exposed to extremely high levels of dioxin for some time with no adverse affects, except the skin rash.

In her book, *Toxic Terror*, Dr. Elizabeth Whelan sums up these kinds of scare tactics rather well when she writes, "There is nothing resembling a significant number of scientists and public health officials who feel that the dioxin and dioxin-containing herbicides have been responsible for the crimes with which they have been charged. To the television news programs, however, the scientific community's consensus is offered by a select group of the same 'experts' over and over again.

"Rarely do we see coverage about the expert who finds, for example, that dioxin in soil samples does not pose much of a health threat to humans in the area. What we do see on television news program are interviews with the same few scientists preaching horror stories and doomsday prophecies. During early 1983, almost every television report, documentary, and print story on dioxins cited the University of Illinois' Dr. Samuel Epstein and Harvard's Dr. Matthew Meselson. Epstein, in particular, has frequently been quoted in absolute terms when the data on which he bases his statement are not accepted by the scientific community, such as this excerpt from the Washington Post: 'The evidence is overwhelming that dioxin is carcinogenic in humans.' The news media want sensationalism, and they know where they can get it." Actually, there is no evidence that any of the various chemicals called dioxin are carcinogenic in humans.

Most of the money for toxic cleanup through the superfund is aimed at dioxin cleanup. Most of these sites have extremely minute amounts of some form of dioxin measured in parts per billion or even parts per trillion. That's like one drop in a very large lake. There were people in Italy who were *covered* with up to *four pounds* of the stuff and suffered no more than a temporary skin rash. Decades later, there are no signs of any long term effects. There are more than a thousand sites listed for cleanup by the superfund. So far, about a dozen have been cleaned up at a cost of over *nine billion dollars,* or almost a billion dollars a site. Have we lost our minds or just all common sense? Do you know what nine billion dollars could do for education if properly spent? The total cost for cleaning up these extremely minute amounts of dioxin may be as high as ten trillion dollars. That's over two times our total national debt which, by itself, will soon cause our economy to crash.

It gets even sillier. Much of the dioxin contamination comes from the air and is produced by volcanoes, forest fires, and auto exhaust. It would make more sense for Congress to outlaw volcanoes, forest fires and automobiles than what we are doing now. Anyone want to try to stop the EPA's multi-trillion dollar juggernaut before they bankrupt us?

OZONE

The ozone layer is self-healing. It is regenerated by the sun. It reached its lowest recorded level in the winter of 1992 - 1993. As of October 1994, the ozone level was back above normal. This is not due to the infinitesimal reduction in man-made/released fluorocarbons (CFCs), but to the fact that we have not had a major volcanic eruption recently and that variations in the sun's activity have vastly more affect on the ozone levels than anything man can do. The rise in ozone levels due mainly to the lack of any major volcano eruptions, which causes, by far, the most reduction, is a clear verification that the ozone is constantly being renewed. The CFC restriction nonsense for cars is a political/economic ploy that will have no effect on the atmosphere, but a considerable effect on people's pocketbooks.

The ozone "hole" that appears annually over Antarctica is not a hole at all, but a reduction in the amount of ozone. The hole is caused by extreme cold and the annual Polar Vortex, which is a strong, cyclone-like storm. It has nothing to do with man-made CFCs, but apparently Congress doesn't know that. Even worse, there is strong evidence that all of the ground based measurements of ozone depletion are skewed by the amount of sulphur dioxide in the atmosphere, mostly from volcanos. When the measurements are corrected for the sulphur dioxide, *all* ozone depletion disappears from the measurements.

Most ozone depletion, if there is any, is due to chlorine compounds and nitrous oxide. One volcano eruption can increase the chloride compounds in the atmosphere as much as 1000 times more than an entire year's production of chlorine and fluorocarbon compounds, world wide. Very little of that production escapes into the atmosphere. Three hundred million tons of chlorine reach the atmosphere each year from the evaporation of sea water. The amount of ozone depletion from people working on their own cars' air conditioning systems is not even measurable if it exists at all, which is seriously in doubt. Yet, we have regulations that prevent you and I from buying little twelve ounce cans of R-12 so that we can replace lost gas in our car's air conditioners. Instead of a couple of dollars, we have to pay a certified technician $30-150 to service the air conditioner. There is no such restriction on R-22 which is used in home air conditions and other refrigeration. Why not?

New cars are coming out with air conditioning systems that don't use CFCs as they are to be phased out. The result is that they don't work as well, can't keep the car cool in really hot weather and are more expensive. For what? There isn't the slightest scientific evidence that the ozone would be affected at all if every ounce of man-made CFCs were *released* every year. On top of that, we will soon be unable to buy R-12. When your car loses enough of its gas, you can either buy a new car, replace the entire air conditioning system with the new version, or sweat. This is just another example of a scientifically illiterate Congress over reacting to extremist groups and a bureaucracy running away with power that they should not even have under the Constitution.

There is an old saying that if you want to know who is guilty, follow the money. Who profited from the ban on CFCs? Well, one group is the people who manufacture the new equipment that anyone who services air conditioners has been required to buy. Take a look at which states they are in and who sponsored the legislation that brought this about. As Franklin Roosevelt indicated, in politics, nothing happens by accident.

The ozone scare has been used by a large number of groups, including NASA, to further their agenda. When NASA's budget was up for consideration, and there was a threat of research funds being cut, suddenly a NASA report warned of the probability of a

The Sky is Falling - 151

dangerous ozone hole over the eastern U.S. that summer. NASA pled their case for more research funds to combat this kind of deadly danger. They got the funds. The hole never showed up, and in fact, it was not even possible for it to have shown up.

In past ages, volcanic activity was much higher. The ozone levels were not seriously depleted for very long. There was no known affect on the life forms from this ozone depletion. The ozone scare is just another attack on our constitutional freedoms.

In reference to global warming and ozone depletion, a French scientist said,

> *"Global warming is an outright invention. It is absolutely unproven, and in my view is a lie. A lie that will cost billions of dollars annually.... There is no danger from CFCs to the ozone layer, nor is there any danger from CO_2, no greenhouse effect, nor any risk of any kind of global warming. It is to me, a pure falsehood."* - Haroun Tazieff, volcanologist, geologist, and former Secretary of State for Prevention of Natural Disasters, French Government.

We are probably entering a period in which the earth will cool down somewhat, rather than warm up. The truth is we just don't know. The earth has gone through some drastic changes in climate before without man having any influence on it whatsoever. Will increased carbon dioxide really cause it to warm up? We don't know. What we do know is that increased carbon dioxide causes plants to flourish, which increases the oxygen levels.

There is one area we should be concerned with for a number of reasons, and that is the destruction of the rain forests. These reasons include oxygen production, carbon dioxide reduction, and the protection of valuable plant and animal species. These unique plants may provide many kinds of new medicines. We haven't even scratched the surface of what might be there.

The Communists have changed from red to green, like chameleons. They have heavily infiltrated the environmental movement. Gorby is one of the leading figures in environmental activism from his plush quarters in California that were built for him even before he was "reassigned". To hear him spouting pleas for peace after invading Afghanistan is as ridiculous as him spouting environmentalism when Soviet Communism has left much of the planet incredibly polluted, resulting in serious health effects, even death, for tens of millions of people. While communism takes its one step backwards, according to Lenin's plans, the Communists/Socialists are using this time to advance their agenda of a one-world government through the environmental movement while waiting for their next two steps forward.

It will take a great deal of discernment for us to know which issues are real and need to be dealt with and which are phony or part of some extremist agenda. This kind of discernment is nearly absent from a government made up mostly of lawyers. We need to retire most of the lawyers in Congress and state legislatures, and replace them with engineers, housewives, cab drivers, chiropractors, small businessmen, and other people who are grounded in some kind of reality. Lawyers love laws. Laws are the reason for their existence. That's why we are drowning in them. We can't constitutionally ban lawyers from running for office, but we can refuse to vote for them if given any alternative. Why don't you run for office at some level?

ASBESTOS

Asbestos abatement is another example of hysteria being deliberately generated for political purposes with extremely costly results. Asbestos comes in two basic forms. One is the long fiber, which is rather soft and flexible. The other is a very short, stiff form. The long form is mined and used mostly in the Western Hemisphere. The short form comes mostly from South Africa. The long form is virtually harmless. A lot of people breathe and drink it from natural sources in certain places with no ill effects.

The short stiff form is quite dangerous as it can penetrate air sacs in the lungs and is very hard to dislodge. During World War II, the short form was imported from South Africa and used mostly in shipbuilding. The people who worked with this stuff are the ones who are showing up with the problems.

The asbestos panic was started by Joseph Califano when he was Secretary of the Department of Health, Education and Welfare. After he used a draft report from the National Cancer Institute and the National Institute of Environmental Health Sciences in a speech at an AFL-CIO national conference the political pressure was on. The information in that draft report was self-serving, to say the least. No one would ever admit to authoring the report. Scientists the world over have condemned it as a gross exaggeration or nonsense. Yet, we still have a very expensive asbestos abatement program that is doing far more harm than good.

Asbestos is supposed to be removed from all of our schools. Sounds reasonable, right? Wrong! Measurement of airborne asbestos in schools reveals that airborne asbestos is virtually non-existent, only about 0.0001 fiber per cubic centimeter of air. After "removal" by the best available methods, the airborne asbestos measures *forty thousand* times higher and will remain so for years. Isn't that great progress at enormous costs? The Manville company went bankrupt because of this politically-generated panic.

Asbestos is to be phased out of production with the loss of thousands of jobs. In many uses, such as fire protection suits for firemen, there is no substitute. You and I are paying for raising the amount of asbestos our children are breathing by a factor of 40,000 with this abatement program. If people were really worried about the asbestos that is used to insulate pipes and for other uses, all they have to do is spray them with used motor oil. The gooey oil soaks in and coats the fibers so they stay put. Will OSHA allow such a procedure? Of course not. It is too simple and too cheap. The asbestos that is in use here is not that dangerous in the first place, and the "removal" increases the airborne amount by 40,000 times, but the incredibly expensive abatement program just keeps rolling on.

Acid rain, global warming, and many other issues that are being used to increase governmental control over our lives are likewise being manufactured or exaggerated. Whole books have been written, from solid scientific data, refuting much of what we are being told by the media and the government.

Even people like liberal Congressman John Dingell are aware of how far off base things have gotten. Dingell says:

> *"It is increasingly apparent that there is something fundamentally wrong with much of the science underlying our environmental health regulations, as we have seen in*

recent episodes on asbestos, dioxin, and PCBs, where the risks have been dramatically overstated at simply immense cost to the public... You take the standard on CO. That came out of the work of a VA scientist, who, it turns out, cooked the books. It was criminally fraudulent work. Yet today's clean air standards are still bottomed on his work."

ELECTRICAL POWER

Another area I do want to touch on is electrical power. We are extremely dependent on electrical power. Our power plants are aging, and plans for replacement are lagging way behind demand. If we replaced every light bulb, electric motor and other device with currently available high-efficiency units, we could cut our power use in half. Unfortunately, we do not have the capital available to do this, thanks in large part to the government's rapacious appetite for money for pork-barrel and income redistribution schemes. Given the horrendous expense of building power plants, even fossil fuel plants, due largely to environmental and other government regulations, replacement plants are being delayed or canceled.

The anti-nuke crowd doesn't seem to understand that a fossil fuel plant releases more radioactive material into the atmosphere in one year than a nuclear plant would over its entire lifetime, maybe several lifetimes. That doesn't even count all of the other noxious stuff it spews forth that has an adverse affect on many people's health. Yet nuclear power plant construction has stopped. With the new designs available, these types of power plants are completely fail safe from explosions or meltdown, no matter what happens. In the past, every plant was designed from scratch and subjected to incredible regulation and red tape. By standardizing on a given design, we could build totally safe, environmentally friendly nuclear power plants that are much cleaner and cheaper than fossil fuel plants and really help clean up the air. Will we? Of course not. That would make too much sense and the few anti-nuke noise makers have far more political clout because they are organized than the millions of citizens who have to pay needlessly high power bills and breathe the polluted air.

> *"Let's face it. We don't want safe nuclear power plants, we want no nuclear power plants."* - A spokesman for the Government Accountability Project.

> *"Scientists who work for the nuclear power or nuclear energy have sold their soul to the devil. They are either dumb, stupid, or highly compromised....Capitalism is destroying the earth. Cuba is a wonderful country. What Castro's done is superb."* - Helen Caldicott speaking for the Union of Concerned Scientists

Apparently Ms. Caldicott has never seen the environmental disasters all across the Soviet Union.

The endangered species laws, the wetlands laws, and environmental laws are deliberately being used to destroy private property rights and free enterprise. Surprising as it might seem, the CFR/TLC crowd do not want free enterprise. They want controlled enterprise and trade with them in control.

ANIMAL RIGHTS

> *"ARFs (animal rights fanatics) aren't really concerned with the 'rights' of animals - or trees or rocks or clouds or mushrooms or rotting logs; all those imaginary entitlements are merely a way to attack your rights, especially your right to private property. To do that, major portions of the U.S. Constitution must be repealed or invalidated, and this is the basis of the ongoing frontal assault on animal use, hunting, and the Second Amendment. With the Second neutralized, the government control freaks can have their way."* - Outdoor writer John Wootters

> *"About 130 habitat conservation plans are being prepared on the use of millions of acres of private,*

> *municipal and state land across the country .. the land is virtually unalterable while local governments and landowners struggle to meet federal mandates."* - Leslie Spencer, *Forbes Magazine*

In Idaho, a group of people are fighting the "endangered" classification of a pin-head-sized snail that differs from other snails only in that it has a larger-than-normal sexual organ. There are hundreds of such "species" on the list that the environmental fanatics want listed so that they can control how your (or someone else's) property can be used.

I believe in trying to protect truly endangered species within reason. Unfortunately, we have lost more species in the animal kingdom, without any assistance from man, than now exist. We will continue to lose them. Reasonable measures to protect all of the remaining species that we can makes sense for several reasons. Insanity in pursuit of impossible goals (or political agendas) at incredible costs does not.

The U.S. Fish and Wildlife Service plans to declare 800,000 acres in Texas as "critical habitat" (now a legal term) for a migratory songbird called the golden-cheeked warbler. This will make the land virtually unusable for anything. A lot of people are fed up with this sort of thing and are starting to organize groups and protests. There isn't any real evidence that this sort of restriction of land use will help these birds one bit. It is all speculation.

The bureaucrats are using these laws in ways that Congress never intended. Often, the laws don't even make sense. David Trotter led a group that wanted to develop about a thousand acres into residential housing. Fish and Wildlife officials agreed, but they insisted he hand over to them over 700 acres of the land as habitat for the golden-cheeked warbler and some cave bugs. Then they demanded that the group buy approximately 900 more acres and donate that for additional fish and wildlife habitat, to mitigate the harm they were doing(?) by building homes on the 340 acres that they had left, only one-third of the land that they wanted to develop. In other words, to get permission to build houses on 340 acres, not the original thousand, they had to give 1600 acres to the Fish and Wildlife Service. If you or I tried to do something like what the

The Sky is Falling - 157

Fish and Wildlife Service is doing, we would go to jail for extortion, which is precisely what it is.

If there is a valid reason to protect some property for important public reasons, then let the government buy it – rather than stealing it, extorting it, or determining its use. Private property rights, along with the right to a trial by jury, and the right of self defense, are the foundation of all other rights and they are under relentless attack by groups who want to run the world according to their particular agendas. Unless Congress puts strict limits on the government's authority (fat chance), bureaucrats will continue to run off of the cliff with it. Senator Mark Hatfield was one of the authors of the original Endangered Species Act and has this to say about the way it is being misused:

> *"..it has come to be a law that favors preservation over conservation. There is no question that the act is being applied in a manner far beyond what any of us envisioned* (as usual-RT) *when we wrote it twenty years ago....But today the act is being applied across entire states and regions, with the result that it now affects millions upon millions of acres of publicly and privately held land, and many thousands of human beings....The situation has gotten out of hand."*

Indeed it has gotten out of hand. Here is just a small sample of the costs from Dixie Lee's book, *Environmental Overkill.*

1. Three billion dollars have been spent to protect housewives from labels that proclaim spaghetti sauce to be fresh. (Nanny government strikes again-RT)

2. The rehabilitation of 222 sea otters was mandated after the Exxon Valdez oil spill at a cost of more than $80,000 *per animal*. (There are thousands of sea otters in the area-RT)

3. The Stevens Kangaroo rat recently received exclusive rights to land worth $100 million. (Did the government buy the land? Of course not.-RT)

4. The regulatory bureaucracy drains $13.5 million from the economy for each premature death averted by a rule governing arsenic emissions from glass plants, or $5.76 TRILLION per premature death averted by a regulation covering wood-preserving chemicals. (Never mind; it's only your money! It's no wonder we are losing our jobs to other countries and our ability to compete. Why would any multinational corporation in its right mind stay in this country?-RT)

5. Dr. J. Laurence Kulp calculates the cost of the acid rain requirements of the 1990 Clean Air Act at $4 billion a year. The benefits come in at just $100 million. (No wonder the environmental extremists claim that it is unlawful to take possible benefits, if any, into account. The government should just spend your tax money because these people say so.-RT)

The Joint Economic Committee for the U.S. Congress says that the costs of administering and policing all federal regulations was $500 billion dollars for 1992. It has gotten a lot worse since then. The budget for 1993 includes $562 million for implementing federal regulations. That figure is *double* the defense budget that the Democrats always want to cut. Many of these regulations are an unconstitutional power grab by the federal government. Let's get rid of those that are not clearly authorized by the Constitution and save about two-thirds of this money. This includes only the money spent by the federal government. This does not include money spent by state or local governments and the costs that businesses incur in dealing with all of these regulations. The cost to businesses is probably three times what the government spends. It's no wonder we buy most of our things from other countries. The government has hidden some of the results of all this by deliberately lowering the value of the dollar, to try to remain competitive with the effect of drastically worsening our balance of payments deficits and making us all poorer.

If you want evidence of the real purpose behind most of these laws, the National Biological Survey provides a classic example. This was supposed to be an innocuous survey, at great expense, to determine what we really had in terms of biological resources. It was, in reality, no such thing. It was a blatant attempt to destroy private property rights. When amendments were added to the act to protect private property rights from government trampling, the Clinton administration suddenly dropped their support for it. Guess why?

The attitudes and beliefs of many of the people driving these agendas was demonstrated quite clearly by Peter Berle, President of the Audubon Society, who simply dismisses the idea of property rights. They don't exist as far as he is concerned. Similarly, Michael McCloskey, Chairman of the Sierra Club said, "Trees and rocks have rights to their own freedom." In other words, trees and rocks have rights, but you and I don't, even though ours are clearly spelled out in the Constitution.

Groups like the Audubon Society and the Sierra Club have done a lot of good in the past. It appears that they have been captured by the lunatic fringe and are often being used for purposes far removed from what their supporters are told. They have elevated the use of government agencies and our tax money to advance their agendas to an art form. Trying to fine-tune the system to prevent some of these flagrant abuses simply won't work. We need to return to having the federal government do and fund *only* what the Constitution allows it to do. That would solve most of the abuses rather quickly.

Many of the self-appointed leaders in the environmental movement want to cut the population of the world in half, as quickly as possible. They think that people are the problem. If you wonder why I put that quote about the extinction of the human species at the beginning of this chapter, it was just to set the tone of what we are up against. I mentioned Gorbechev as the leading environmentalist spokesman for communism, but many other very powerful people share most of these extremist views.

> *"If I were to be reincarnated, I would wish to return as a killer virus to lower human population levels."* - Prince Phillip–who is also the leader of the World Wildlife Fund.

> *"If radical environmentalists were to invent a disease to bring human populations back to sanity, it would probably be something like AIDS. It has the potential to end industrialism, which is the main force behind the environmental crisis."* - Earth First newsletter.

These environmental wackos believe the state (U.N.) should strictly control birthrates, like China does.

> *"The right to have children should be a marketable commodity, bought and traded by individuals but absolutely limited by the state."* - Kenneth Boulding, originator of the Spaceship Earth concept.

In other words, only the wealty elite (CFR) should be allowed to have as many children as they want.

> *"Childbearing ..a punishable crime against society, unless the parents hold a government license....All potential parents ..required to use contraceptive chemicals, the government issuing antidotes to citizens for childbearing."* - David Brower

The probable negative affects of these chemicals on the whole population are irrelevant. They want to get rid of people anyway. These people continually invent or distort potential problems to try to get total control of the earth into their hands.

> *"Above all, we need to learn to act decisively to forestall predicted perils, even while knowing that they may never materialize. We must take action, in a manner of speaking, to preserve our ignorance. There are perils that we can be certain of avoiding only at the cost of never knowing with certainty that they were real."* - Jonathan Shell, author of *Our Fragile Earth*.

> *"We have to offer up scary scenarios, make simplified, dramatic statements, and make little mention of any doubts we may have. Each of us has to decide what the right balance is between being effective and being honest."* - Stephen Schneider, who dreamed up the theory that CFCs were depleting the ozone.

In other words, just like Lenin, lie to advance the cause when needed. The real objective of many of these people is to return us to a Stone Age existence.

> *"We have wished, we ecofreaks, for a disaster or for a social change to come and bomb us into the Stone Age, where we might live like Indians in our valley, with our localism, our appropriate technology, our gardens our homemade religion – guilt-free at last."* - Stewart Brand, *Whole Earth Catalogue*.

If these people were just writing offbeat newsletters and books, it wouldn't be of serious concern. However, Vice-president Gore is one of the leaders as are many people around the globe with very powerful political influence. They are determined to control everything you and I do, and use our tax money and the power of Big Brother government to do it.

Currently, the Republicans are trying to eliminate many of the environmental laws, partially in response to the excesses that have occurred. Some of that is needed. But, as usual, Congress is overreacting in some areas. The federal government has a proper constitutional role to play in fighting pollution that crosses state lines and in protecting wildlife that does likewise. The authority for federal protection of isolated, non-migratory wildlife is not in the Constitution, but reasonable approaches can be developed in spite of this. The primary reason for gutting many of the environmental laws comes from the money that these politicians receive from huge corporations and PACs. These laws hurt profits and competitiveness, so now the pendulum is swinging back the other way to excess.

We need to find honest, knowledgeable people to pass laws that address real problems in reasonable ways and stop the bureaucrats from running away with their unconstitutional authority and catering to the lunatic fringe. Anyone know where we can find about six hundred such people? Don't bother looking in Washington.

Chapter Seven

The Russians are Coming

> *"War to the hilt between communism and capitalism is inevitable. Today, of course, we are not strong enough to attack. Our time will come... To win, we shall need the element of surprise.* The bourgeoisie will have to be put to sleep. So-we shall begin by launching the most spectacular peace movement on record.** There will be electrifying overtures and unheard of concessions.*** The capitalist countries, stupid and decadent, will rejoice to cooperate in their own destruction. **** They will leap at ANOTHER***** chance to be friends. As soon as their guard is down, we will smash them with our clenched fist."* - Dimitri Manuilsky, speaking at the Lenin School for Political Warfare in the 1930s

[COMMENTS-RT: *It would not have been any surprise if the Soviets had attacked anybody in the West before now. The surprise is yet to come. **Gorby and Yeltsin are preaching peace everywhere they can. *** As in giving up captive nations and opening up files and records. **** As in financing the modernization of the Red army. ***** Here we go again.]

The time has come for the last perestroika. The one Gorby brought into being is not the first, it is the sixth and most extensive. It follows the same script as the previous ones, only more so. The others were tune-ups. Each time, the Russians pretended to open up to more capitalism. They relaxed their iron control over the people and the press. They "moved" towards democracy. They suckered in massive amounts of Western money to shore up their failing economy while they stole Western technology and built up their military power. They are doing exactly the same thing this time. Remember when the news media was telling us, after Yeltsin took power, that we had to send food and money to get the people through the coming winter? Then and now, the Russians continue their military production, including building nuclear submarines at a cost of three billion dollars each. They are selling weapons all over the world to finance updating their military with stolen Western technology. Each time, when they had gotten as much money and advantage as they could, the crackdown came. It has started now in Chechneya.

The first perestroika lasted from 1921 through 1929 under Lenin. Number two lasted only from 1936 to 1937 under Stalin. He had a hard time letting anything loosen up, and primarily used it to eliminate his enemies when they surfaced under the new "freedom". The Chinese communists call it "letting a thousand flowers bloom." Then the flowers are easier to find and rip up by the roots. Number Three lasted from 1941 to 1945 under Stalin because of the war. Once Germany was defeated, Stalin seized most of Eastern Europe. Number four was from 1956 to 1959 under Kruschev. Number five came during 1970 to 1975 under Breshnev.

The present one started by Gorbachev and carried on by Yeltsin is starting to shut down as the "reclaiming" of the former "states" begins. Could this whole "breakdown" of the Soviet Union really be a ruse? Read the quote at the beginning of this chapter. Remember, they are the world's master chess players. They always plan a dozen moves ahead. Russia did not grow from a small, insignificant principality around Moscow in 1550 to become the largest empire the world has ever known by accident. Rudyard Kipling had them pegged when he wrote a poem about how the Russian Bear who having been wounded, stood up as in supplication. The hunter, moved by pity, moved toward the bear, who reached out and ripped his face off.

> *"When he stands up as pleading in wavering man brute guise,*
> *When he veils the hate and cunning of his little swinish eyes,*
> *When he shows as seeking quarter, with paws like hands in prayer,*
> *That is the time of peril – the time of the Truce of The Bear."* - Rudyard Kipling

The Russians are Coming

Welcome to the time of the truce of the bear. Those who will not learn from history are condemned to repeat it. Am I saying all the breakdown of the Soviet Union and the reuniting of Germany was planned? You bet!. The "coup" was as phony as a three-dollar bill. The Communists are masters at coups. They have been pulling them off for more seventy-five years. Only the most naive would believe the inept, bungling image the Soviets have been deliberately projecting. In how many failed attempts against any communist government did the perpetrators survive? One. The phoney one. Remember, everyone involved in that show was handpicked by Gorbachev. Where are they now, dead or in prison for life? Hardly. They are all *back in power*, except the one who didn't want to go along with the charade in the first place. Were you aware that the troops surrounding Yeltsin at the Russian White House were, in fact, loyal to Yeltsin?

How many coups have the Communists staged where the means of communications were not seized and silenced? Only one. That is always the first target after seizing the leaders. Yet throughout the whole play, even the Western media had open access to broadcast the whole show for our consumption. How many communist coups have you known about where the objects of the coups (Gorby and then Yeltsin) were surrounded by army troops and survived? One. If you still think the coup was genuine, please explain why the incorporation papers for Gorby's Foundation in California were filed four months before the "surprise" coup, and everything was in place in California ready for him when it came time for him to exit stage right. His current position allows him to control much of the world's environmental movement, which the "former" communists have heavily infiltrated, and to maintain close ties with the New World Order leaders. Question, who is financing his "foundation"? Three guesses! It is the same foundations whose board members are also members of the CFR/TLC. He is also a leader in the "peace" movement and the convergence of communism and capitalism.

> *"You cannot move directly from capitalism to communism. Socialism is a necessary stepping stone to*

communism and hence all communists should work for socialism." - John Strachey, British Labor Party

Gorbachev is a much more powerful leader than Yeltsin in the whole scheme of things. Yeltsin will soon be written out of the script. Gorby won't. While the communists focus more on using the environmental movement to suppress national sovereignty, the capitalistic New World Order types are concentrating more on using treaties like GATT, NAFTA and Rights of the Child to do it. They are cooperating closely and supporting one another in all such endeavors. Isn't it interesting how the "former" Communists are suddenly so interested in global control of the environment for everyone else, when their countries are, and will continue to be environmental disaster areas?

How do we know this is all a giant play on the world's stage? One way is that we had the script well in advance. In 1980, Anitoliy Golitsyn, who is one of the few real defectors from the KGB and not a double agent, wrote a book called *New Lies for Old*. He was ignored by our government. For years before Perestroika Six even started, he told them what was planned. In his book, written in 1980, he detailed the entire script from official Soviet sources that we are now witnessing being played out to the letter. The script goes as follows:

1. Free the captive nations of Eastern Europe, for the time being. Subsidizing most of these basket cases was draining the Soviet economy which itself was faltering. Let the West come in and supply the needed capital to get them on their feet, then take them back. (How many of these countries still have Soviet troops inside their borders? Eighteen of the twenty-two countries that came from the breakup of the Soviet Union have communist leaders back in power.)

2. Remove the Iron Curtain and tear down the Berlin Wall as a gesture of their "conversion" to democracy and openness. (More deception.)

3. Support German reunification as a condition to moving Germany to a more neutral position and helping undermine NATO.

4. Purge the old communist leaders in Eastern Europe, and bring in their chosen reformers under new names. Declare communism dead and adopt new terminology and labels.

5. Set up supposedly free elections. Establish opposition parties that the KGB controls to give the appearance of democracy. (How many elections recently have put "former" communists right back into control of not only Eastern Europe, but all 15 Soviet Republics? Some reform! Can you really believe that all these countries are actually electing their former communist masters to rule over them again of their own free will? Our media seems to believe it)

6. Use the desire of the West to be friends and the new spirit of openness and cooperation to infiltrate thousands more KGB agents into the west, particularly the US. Streamline and merge the various intelligence services. Reorganize them to better penetrate the West. (More deceptive name changes.)

The goals of the communists for all this elaborate deception are as follows.

1. The neutralization or disbanding of NATO. (This has been a primary goal of the Soviets since it was formed, and they will sacrifice almost anything to accomplish it. Already many of the NATO operations have been transferred to the Organization on Security and Cooperation in Europe, which Russia effectively controls. American troop strength in Europe has been reduced from 400,000 when all this started to 100,000 today and soon will be down to zero. This long-cherished goal is almost within reach.)

2. U.S. troops will be pulled out of the Philippines, Korea, and other areas under threat of communist guerrillas or armies. (Bush obligingly pulled our troops out of the Philippines, and if North Korea will simmer down, we will soon be out of Korea.)

3. The West (as in American taxpayers) would assume responsibility for the $50 billion Soviet debt. (The Russians have

done a masterful job of the good cop [Yeltsin] bad cop [Zhirinovsky] routine. We "have to" keep pouring our money into the Soviet empire to make sure Yeltsin [democracy?] prevails and Zhirinovsky [fascist] doesn't. Sure, we do. Zhirinovsky or whoever the KGB has waiting in the wings will take over precisely when they want him to do so. By the way, we have pretty well taken over the cost of the Soviet debt. Chalk up another resounding disaster for our politicians.)

4. Merge the Western and Eastern Europe countries into one giant trading block. (Sound familiar? Russia plans on eventually dominating this trading block because of its immense natural resources, cheap [slave] labor, and military power.)

The above strategies and goals are only a small part of the Russian plans for you and the rest of the world. If you want all the gruesome details, get a copy of Golitsyn's book. Remember, it was written in 1980 before any of this started, and taken from official Soviet documents. He has just released a second book called *The Perestroika Deception*. It is based on the memos he has written to the CIA for the past ten years, warning them about the coming perestroika number six. Apparently, they ignored the memos.

The Russians believe that the West will be in such bad shape within ten to fifteen years, economically and morally, that they can spring the waiting trap and take control with the help of their New World Order partners. The Gramsci strategy detailed elsewhere has gone a long way toward destroying the moral values of the West already, in cooperation with their humanist collaborators. It is interesting to watch these two groups try to use and play off of one another. Mayer Chapel Rothschild said, "Permit me to issue and control the money of a nation, and I care not who makes its laws." We gave control of our nation's money to these international bankers in 1913. They are moving toward their New World Order controlled by the international banks and corporations on this philosophy. On the other hand, the Communists are pursuing Mao's philosophy that political power comes out the end of a gun barrel. Each group thinks that they can use and control the other. Remember the opening quote in the beginning of this chapter. Would you care to bet on which group comes out on top in the end? I would bet on the guns. Russia will be able to survive the coming economic meltdown better than the West will. Most of us will be clamoring for the government to do something, and will sell our birthrights for a bowl of oatmeal.

> *"The capitalists of the world and their governments, in the pursuit of the conquest of the Soviet market, will close their eyes to the indicated higher reality, and thus will turn into deaf, mute, blind men. They will extend credits in giving us the materials and technology we lack. They will RESTORE our military industry, indispensable for our future victorious attacks on our suppliers. In other worlds, they will labor for the preparation for their own suicide."* - Vladimir Lenin

Does that sound like today's news stories? We are following Lenin's script to the letter. We are doing it again. I could write a whole chapter on how we have been subsidizing the Soviet military machine for years and are *still* doing it, even more so now.

The Russians signed a secret treaty with Germany several years ago. It established which areas would be under German "influence" or Russian "influence". In return for a free hand in the areas of historically German hegemony, Germany would not press for the removal of Russian troops from the former slave states or interfere in Russian "peacekeeping" in its former territories. This set the Serbs free to pursue their ethnic cleansing with Russian support. If you wonder why we can't seem to do anything about the situation in Bosnia, you might look to that treaty. This is Bismarckian realpolitik at its finest.

If you think that all those Soviet nuclear bombs and missiles are really out of control of the KGB, you are very naive. It is part and parcel of the communist disinformation campaign. The KGB knows exactly where those nuclear devices are, where they are going and why. Always remember who has the codes to arm those things. Those nuclear experts going to other countries are an excellent way to earn badly needed hard currency and they also allow the KGB to keep very close tabs on what these other countries are doing. The main reason for the nuclear scare tactics (terrorists with nukes) is to speed arms control and disarmament, while the Russians quietly build up and modernize their military power. They have violated every treaty they ever signed. They still are doing it today.

Remember, the Marxist/Leninist dialectic is two steps forwards, one step backwards. The Communists have just taken one giant step backwards, and are now beginning to take two giant steps forward. Russia did not expand from a small patch of ground around Moscow in 1550 to build the largest empire the world has ever known without having a lot of patience, perseverance and ruthless determination. They were not and are not the bumbling buffoons and morons that they currently are projecting to the world. Those who will not learn from history will..... oh, never mind.

> *"The point is that the communist goal is fixed and changeless-it never varies one iota from their objective of world domination, but if we judge them by the direction in which they seem to be going, we shall be deceived."* - Mrs. Elena Sakharov

> *"Russian leaders do not conceal their interest in war. In speaking of former Soviet Republics many no longer use the phrase 'near abroad' but the much stronger 'common defense space.' Their Foreign Intelligence Service cites Russia's two 'crisis zones.' These are Central Asia and the Caucasus, both areas where Russians and Moslems are now fighting. Yeltsin has ordered his generals to train their troops for 'local conflicts'. In a July meeting about the war between Christians and Moslems in the Balkans, Foreign Minister Andrei Kozyrev spoke openly about a possible World War III."* - Richard Maybury, *Early Warning Report*

The Institute for Soviet Studies has stated that Russian armed forces are twice as large as official statistics indicate. Russia has 4.8 million men under arms, not 2.3 million.

> *"As I grow older, I pay less attention to what men say. I just watch what they do."* - Andrew Carnegie

Don't listen to the well orchestrated disinformation campaign on the evening news. Watch what the Russians are actually doing as in Chechneya. The Russians are playing a very intricate game of geopolitics. They are wooing the fundamentalist Moslems in Iran, while supporting the Serbs in getting rid of the Moslems in the former Yugoslavia. They are reasserting control over the Moslem states they once ruled to try to prevent the spread of fundamentalism northward and to maintain control of the areas they need for their eventual invasion of the Middle East, particularly Israel.

Their what? The Russians have been planing to invade the Middle East for decades. When the Israeli army invaded southern Lebanon some years ago, they found an astounding cache of weapons and other support material for a massive invasion. It was far beyond what a few Arab guerrillas could use in a hundred years. The Russians haven't changed their goals one bit. With the deterioration of their own oil and mineral industries, they will need to push up their timetable some. Eventually they have to invade the Middle East for several reasons. They have been planning on doing so for centuries. In gaining control of South Africa's strategically essential mineral resources, which they are now consolidating, they have begun their strangulation strategy to bring down the Western economies by controlling access to vital resources. Before the rigged elections in South Africa even took place, the Russians were meeting with Mandela's people to determine how to best "allocate" the resources.

> *"We will take the two treasure chests upon which the West depends, the strategic oil reserves in the Middle East and the strategic minerals in South Africa, and then we will dictate the terms of surrender to the United States and to the West."* - Leonord Brezhnev.

They already have South Africa, thanks to the CFR. The Middle East is next. As for South Africa, it was the New World Order crowd that brought you the current communist government in S.A. Botha, who betrayed S.A. into the communists's hands, is a member of the CFR. Apartheid, which was a disastrous and unconscionable policy to start with, was not the real issue for the people pulling the strings. It was control of the resources. Neither the Communists nor the New World Order crowd care two cents about apartheid one way or another, since we are all to be slaves eventually. It was a convenient excuse to bring the Communists to power. The government should have gone to the much more numerous and democracy-loving Zulus and their allies. Fat chance. Just watch what happens in S.A in the next few years, especially after Mandella is gone. He knows how badly western capital (here we go again) is needed to upgrade the standard of living and develop the resources. He needs that to keep the lid on. So far, he is putting on a pretty good show. When it comes time for the New World Order to take control, watch what happens to the availability of those vital resources. Russia and S.A. produce virtually all of the world's chromium, and much of the other minerals we need to survive in a high-tech world economy. Remember the golden rule. "He who has the gold makes the rules."

I mentioned Chechneya several paragraphs ago. In case you have some questions about that operation, since it doesn't seem to fit the image Russia usually tries to portray, I will give you my opinion, since you asked for it. (You bought the book.) Why did Russia invade this "sovereign country" when such a move would be so unpopular with the Russian people? First of all, who cares what the Russian people think? The KGB, or whatever their name is this week, the army and the nomenklatura are still firmly entrenched and very much in control. Don't believe the propaganda about Russia being a democracy, far from it. As Alexandr Solzhenitsyn said on September 2, 1992, "Communism has by no means collapsed... Don't forget through all this, the structure of the KGB has been retained. The KGB remains a large force with a large apparatus and long far-reaching tentacles." That is not an outdated quote. It is just as true today.

Second, the Russians cannot afford to allow these countries to have true independence for a number of reasons, some of which were mentioned previously. They had to send a message to other former "republics" that such independence will be severely punished. They are in the process of restructuring the empire according to the model mentioned by Gorbachev in a previous quote.

If that is true, why did they send green untested troops to put down the "independence"? There are two main reasons. They wanted to get these troops the training they were going to need to fight in similar situations elsewhere. They also wanted to portray the Russian army as a toothless and clawless tiger to the West, which they did in full view of the Western media. How many other Soviet battles have you seen on TV? They were willing to take the PR hit for going in, but wanted to get as much mileage as possible out of doing so in convincing the West, particularly Europe, that they had really had nothing to fear from such an inept bunch. Do you remember that General who refused to order his troops to attack? That was great Russian theater. Do you seriously think he did that without permission? If so, he would have been dead a few days later. That's how the Russian army operates. Ask anybody who has been in it.

Did you see all the coverage by the media when the seasoned army troops went in and quickly cleared the city? You didn't? Do you wonder why? Three guesses!

Beware the Russian Bear when he stands up in supplication and friendship.

(A significant part of the information in this chapter and some of the information in several other chapters was found in Don McAlvany's book, *Toward a New World Order*, and several of his newsletters.)

Chapter Eight

How We Got to This Point

"Remember, in politics, nothing happens by accident. If it happens, you can bet it was planned that way." - Franklin D. Roosevelt

The oldest English speaking settlement at Jamestown was founded in order to expand the boundaries of Christianity. The Pilgrims came to America to escape religious persecution by European governments and to freely follow their religious convictions. They were looked upon as the Branch Davidians of their day by their governments, only more so. Thus began the previously unheard of expression of religious freedom which has been part of the genius of the American experiment in freedom and self government. That tradition ended in Waco.

The trivializing, smearing and persecution of the three primary monotheistic religions has been growing in America since Darwin. The attacks accelerated after the Scopes trial and have been more blatant for the past thirty years. So far, this campaign has been aimed primarily at Bible believing Christians, but Moslems and Jews will also be targeted more and more, primarily through the influence of the proponents of New Age religions. We haven't seen anything yet.

In 1776, the insufferable tyranny of King George became too much for many of the colonists and brought about the drafting and signing of the Declaration of Independence. The unique American experiment in individual liberty and self government had begun. In that same year, Mr. Adam Weishaupt founded the Illuminati on May 1. Their purpose was to bring about a one-world, universal, occult government. Their plans were well thought out and included a detailed schedule that was to bring their goals to fruition between 1997 and 2000. These goals are right on schedule. The Illuminati were "the enlightened ones". They were deeply involved in the occult and took most of their philosophy from witchcraft and the ancient Babylonian mystery religions. Thomas Jefferson was well aware of their existence when he said: "Their early high level political influence can be found on our currency in the form of the pyramid, which is a symbol and attractor for occult power and the all seeing or third eye that sits atop the pyramid." George Washington was also well aware of them when he said, "I have heard much of the nefarious and dangerous plan and doctrines of the Illuminati, but never saw the book until you were pleased to send it to me. It was not my intention to doubt the doctrine of the Illuminati had not spread in the United States. On the contrary, no one is more satisfied on this fact then I am." The Illuminati planned and directed the French Revolution. They brought in and paid the hoodlums that stormed the Bastille.

The goals of the Illuminati included abolishing private property, abolishing religion, abolishing monarchies, national governments and borders, abolishing the right of inheritance, abolishing the family and all moral values, and instituting communal education of all children. Is it any wonder the Communists also celebrate May 1 in honor of Weishaupt as well as the Russian revolution?

If you have any doubts about the long-range plans of these people, Dr. Albert Pike, a Luciferian, as well as one of the Illuminati, wrote the following words on August 15, 1871 about how they would start the *Third World War*, long before the First World War was even on the horizon. World War III "would be fomented by political strife as stirred up by agents of the Illuminati between the Zionist and the leaders of the Moslem world. The third great world war's purpose is to cause the nations to fight themselves into a state of complete exhaustion physically, spiritually, mentally and economically." Notice that in 1871, nobody would have believed that Israel would exist as an independent nation again, yet this conflict between the Jews and the Moslems was planned to start World War III.

While the Illuminati, as such, disappeared from public view about 1900, in part because of the information about their activities that was published, they continued to pursue their goals through other organizations and personal networking. Their ceremonies were continued in the initiation rites of the Skull and Bones Society at

Yale, which also promotes their goals of a one-world government. Before the CFR was founded, the international bankers, financiers, and owners of large corporations cooperated in a rather loose manner to bring about their control over business and governments as they moved toward a one-world government.

Nineteenth-Century British financier Cecil Rhodes supported world government. He established the Rhodes scholarship program to promote his beliefs. Many leaders in industry and government have been through his socialistic indoctrination, including President Clinton and many others in the present administration.

Andrew Carnegie was committed to a one-world government and pressed for the establishment of a league of nations with armed forces to enforce the league's decrees. Woodrow Wilson was close friends with Carnegie and others in his circle and secretly promised them that he would enter World War I, even though he was publicly pledging to keep the U.S. out of the war.

Woodrow Wilson knew about these people all too well. In 1913, he said, "The government which was designed for the people has got into the hands of their bosses and their employers, the special interests. An invisible empire has been set up above the forms of democracy." That same year, this "empire" passed the income tax amendment, the Federal Reserve Act and the Seventeenth Amendment providing for the direct election of senators. This last step sounded like it was more democratic. The real purpose was to remove much of the power from the state legislatures and concentrate it in Washington.

The Council on Foreign Relations was founded in 1919 to better coordinate the efforts of those committed to a one-world government. Their goals are being pursued by the CFR, the Trilateral Commission (founded in 1973 by CFR Chairman David Rockefeller), the Bilderbergers, and the Council of Rome. The Trilateral commission focuses on trying to bring about three major world trading blocks: the European Community, North and South America, and the Far East. They know that trying to get a one-world-wide trading block would be impossible directly, so they are planning to form three major trading blocks that then can be merged "for everybody's good". The Council of Rome works primarily to bring about the European Community. The CFR is the largest body with the other groups being subcommittees of a sort. The Bildebergers, in particular, invite influential people to their meetings to get them interested in their philosophies and to plan strategy.

> *"The Council on Foreign Relations is the American branch of a society which originated in England ....believes national boundaries should be obliterated and one-world rule established. The Trilateral Commission is international ....is intended to be the vehicle for multinational consolidation of the commercial and banking interests by seizing control of the political government of the United States."* - Barry Goldwater

Many of these people don't accept the idea that they are part of some dark conspiracy to rule the world. They don't want to rule the world; just control those who do for our own good.

> *"They that govern most make the least noise. You see, when they row in a barge, they that do drudgery work, slash and puff, and sweat. But he that governs sits quietly at the stern, and is scarce seen to stir."* - Selden

They do consider themselves the elite of the world, as in Nietzsche's philosophy. Mostly, they want peace, stability, cooperation among nations, favorable trade conditions, and allocation of the world's resources on a basis favorable to them. To accomplish these things, national sovereignty must be subordinated to the larger goal. Nationalism must be suppressed, and nations must be forced to accept world wide government under threat of force from U.N. armed forces if necessary. Like all true elitists, they want to protect us from ourselves. The question is, who is going to protect us from them? They want a *Pax Romana* to facilitate their business. What they fail to realize is that with this kind of power at stake, given basic human nature, such a world government must eventually end up as a one-man dictatorship just as Rome did.

The CFR consists mostly of international bankers, officers of large corporations, Eastern Establishment intellectuals, and politicians. Many Congressmen and most Presidential nominees from both parties are/were members. Many high level governmental appointees are also members. The CFR usually controls the election of the President through the cooperation (corruption?) of the media, whose owners include many members. Often, both Presidential candidates are members, and are committed to a one-world socialist government, although they usually couch it in terms of "cooperation".

Why are so many of America's leaders part of these groups? According to Norman Dodd:

> *"The careers of men are watched. The men who indicate that they would be especially capable in terms of the aims of this group are approached quietly and invited into the inner circles. They are watched as they carry out assignments and eventually they are drawn into it under circumstances which make it virtually impossible for them to ever get out of it."*

Bill Clinton is a classic case of this approach. I might add that it is easier to get someone involved if that person wants to associate with the world power brokers and wants their help in getting elected or promoted to bigger and better things. For a long list of these individuals who are/were members of the CFR/TLC, as well as the foundations and corporations that support these groups, contact F.R.E.E., listed in the Appendix.

George Bush's thousand points of light are straight out of the occult Illuminati material followed by the Skull and Bones Society, of which he is a member. His drive for a New World Order is also straight out of their occult material. The Gulf War was a test of whether a United Nations force could be brought together to imposed the will of the United Nations upon a nation that was a threat to the emerging New World Order. Interestingly, it was the West who built up Saddam with its money and weapons. George Bush was a member of David Rockefeller's executive committee of the Trilateral Commission. His oil company was funded by the Rockefellers.

Bill Clinton is a member of the CFR and the Trilateral Commission. He is dedicated to a one-world socialist government. He ignored organized labor's strong and correct opposition to both NAFTA and GATT. He alienated one of the main strengths of the Democratic party, jeopardizing the re-election of his own party members in Congress and his own re-election. The reason? He is totally controlled by the CFR. These treaties supersede not only our laws, but, effectively, the Constitution itself and are the most direct and effective way of getting around the freedom and liberty guaranteed by the Constitution.

Another treaty that is even more insidious is the Rights of the Child Treaty. This is a clear and blatant attempt to take away the rights of parents to raise their children according to their own views and will effectively make all children wards of the one-world superstate. Already, other nations are discovering that parents have lost control of their children, such as a couple in France that were arrested for lightly swatting one of their children, which is forbidden under the treaty.

The late 1800s witnessed the ascendence of atheism and humanism that found a rational foundation in the publications of Charles Darwin on evolution. These were rank speculations based on deception from the start. This subject is covered in detail in Chapter Nine. The results of the philosophies built on evolution have been horrendous, unprecedented human suffering and death.

Friedrich Nietzsche, along with Kierkegaard, were probably the major philosophers of the 19th century and laid the foundation for existential philosophy. Nietzsche leaned heavily on Darwin's theories, as did the others. Their philosophy came to be known as the philosophy of despair because it always ended up in meaninglessness. Life ultimately had no meaning and no real purpose or destiny. That is why many of these philosophers ended up in insane asylums or screaming on their deathbeds about the purposelessness of life. Nietzsche's hatred of democracy and Christianity were to have profound effects on history. He developed the idea of "supermen" who were destined to rule over others, which came to its natural fruition in Hitler's Germany since Hitler was a real student of Nietzsche. That same philosophy is alive and well in the CFR/TLC.

Sigmund Freud's writings were grounded in such philosophies, with sexuality as the center of everything. Engels and Marx based their political, social, and economic theories on Darwin with predictable results. Their writings were a reaction against the virtual slavery of the working class in England in their day. They denounced the profit motive that had run amuck and essentially invented Socialism/Communism as a replacement. Their lack of understanding of human nature – the desire for power and control over people's own lives and those of others, the value of the profit motive and need for incentive, motivation, and competition, and the incredible inefficiency of state-owned means of production – led them to a very flawed solution that turned out in practice to be much worse than what they were trying to replace.

Lenin justified his slaughter of dissenters on the basis of these philosophies. If man was just another animal, the individual human life had no particular value, but must be subordinated to political evolution and the creation of the perfect society. Hitler based his "philosophy" squarely on Darwin and Nietzsche as he often stated. He believed absolutely in the survival of the fittest and was convinced that this described the Aryan race. Mao took extermination of dissenters to new heights. Together, these men, along with Stalin, Pol Pot, and others, brutally killed over 100 million people, based on philosophies built on the foundation of evolution.

Darwinism, although extensively refuted by today's scientific evidence, forms the basis of our "modern education" and the liberals' approach to government, economics, and politics. John Dewey, known as the father of modern education, was a strong advocate of evolution, pragmatism, relativism, and atheism. His views resulted in the progressive education movement in the 1930s. He saw the public schools as the way to indoctrinate vulnerable young minds in the new "science" of evolution and the religion of humanism, not to mention the belief in a utopian one-world government. The disaster that many public schools represent today are the inevitable result of a philosophy of education based on false premises and a very idealistic (and false) view of human nature. His ideas are the foundation of the far-left views now promoted by the leaders of public education in the NEA and the federal government.

Our culture has been sold a phony bill of goods. We have been sold on the idea that we can fill up the emptiness of utter meaninglessness if we get enough money to have or do whatever we want to. As people like Howard Hughes and many others could tell us, that just doesn't work. When people have more money than they can spend and the emptiness is still there, they reach for power, influence, and/or recognition to fill the emptiness. That's where many of the leading members of the CFR/TLC are at.

If you tell people that everything is an accident, that life and all of its struggles are meaningless, that there is no ultimate purpose to anything they accomplish, is it any wonder that our kids shoot or stab each other over an insult, or that a majority of people think that it is okay to lie, or that people will do anything to get money or power to try to fill the emptiness of their lives? If you think about it for long, you will find that the fruits of Darwin's fairy tale are bitter indeed, and actually refute the theory of evolution itself as a godless, meaningless Western civilization descends into chaos, war, and savagery. If there is no higher authority to set standards of right and wrong, good and evil, and to provide a purpose to life, then everybody can make their own rules, since man is the highest source of authority, and civilization as we know it will self-destruct.

The Soviet Union is a prime example. Many of the countries, especially Russia, are now asking, even begging, Christian groups to come into their schools and teach Christianity in a desperate attempt to provide some sort of moral authority and a moral foundation for their society before it disintegrates completely. For seventy years, they taught their people that there was no creator, that they were the accidents of random chance, and that the only way to find some meaning in life was to promote "the Cause", Communism. The only problem was that the people knew Communism, Socialism, collectivism, or what ever you want to call it, just didn't work. It is no wonder that alcoholism is a national disease in Russia. It is no wonder that worker productivity is extremely low and that the quality of work is horrendous. Why work hard if life is meaningless anyway and you can never enjoy the fruits of your labor even in the short term?

FUNNY MONEY

Although the CFR wasn't founded as a visible replacement for the Illuminati until 1919, the international bankers had been fighting for a privately owned central bank in the U.S. from the beginning of the Republic. Washington, Adams, Franklin, Jefferson, Jackson, Randolph, and Lincoln were some of their strongest opponents. The story of the opposition to a central bank and the fight to get one has been the subject of several books.

> *"This institution (Bank of the United States) is one of the most deadly hostility existing against the principles and form of our Constitution." "I sincerely believe that banking establishments are more dangerous than standing armies."* - Thomas Jefferson

In 1913, two laws were passed by the banking establishment that were to change the course of American history far more than anyone could imagine. The first of these laws was the Federal Reserve Act, which established the privately owned federal reserve banks. Congress abdicated its constitutional responsibility to control the money of the country by passing a clearly unconstitutional law. Since then, the bankers have been able to manipulate the money supply for their own benefit, as they have come to control, along with the brokerage houses and investment bankers, many of the world's largest corporations.

> *"The mischief springs from the power which the moneyed interest derives from a paper currency which they are able to control, from the multitude of corporations with exclusive privileges which they have succeeded in obtaining... and unless you become more watchful in your States and check this spirit of monopoly and thirst for exclusive privileges you will in the end find that the most important powers of government have been given or bartered away, and the control of your dearest interests has been passed into the hands of these corporations."* - Andrew Jackson.

How We Got to This Point - 183

I realize that your are probably getting bored with all the quotes, but I want you to realize that this ongoing battle with central banks, international banks, financiers and monopolistic corporations is not some recent wild conspiracy theory. These men who are being quoted fought very hard to prevent exactly what we have now in terms of a central bank and the people in organizations like the CFR and TLC.

> *"If the American people ever allow private banks* (Federal Reserve Banks-RT) *to control the issue of their currency, first by INFLATION and then by DEFLATION, the banks and the CORPORATIONS that will grow up around them will deprive the people of all property* (as they did many people, particularly farmers in the depression-RT) *until their children will wake up homeless on the continent their fathers conquered."* - Thomas Jefferson

Later, after the Bank of the United States had been in existence for a while, Jefferson stated in a letter to James Monroe:

> *"We are completely saddled and bridled, and the bank is so firmly mounted on us that we must go where they ill guide."*

If Jefferson thought it was bad then, he should see it now. I'll give you three guesses as to who controls the interest rates and the amount of money in circulation in the U.S. It is not Congress, but privately owned banks.

After the Second Bank of the United States had been in existence for a while, Andrew Jackson insisted that the power to issue and regulate currency be returned to Congress as stated in the Constitution. During his fight to annul the bank's charter, he said to the bankers:

> *"You tell me that if I take the deposits from the bank and annul its charter I shall ruin ten thousand families.*

> *That may be true, gentlemen, but that is your sin! Should I let you go on, you will ruin fifty thousand families and that would be my sin! You are a den of vipers and thieves. I have determined to rout you out."*

Andy Jackson certainly had his faults, but we sure need someone like him in the White House today to straighten out the financial mess the Federal Reserve System has brought about.

The Rothschilds were behind the push to establish a central bank is this country after the Civil War. This quote is from a letter from the Rothschild brothers in London to some bankers in the U.S.:

> *"...the old plan, of state banks is so unpopular, the new SCHEME will by contrast, be most favorably regarded, notwithstanding the fact that it gives National Banks an almost absolute control of the national finance. The few who can understand the system will either be so interested in its profits, or so dependent on its favors,* (both Republicans and Democrats-RT) *that there will be no opposition from that class, while on the other hand, the great body of people, mentally incapable of comprehending the tremendous advantages that capital derives from the system, will bear its burdens* (government deficits, national debt, soaring interest costs and corrupt politicians-RT) *without complaint and perhaps without even suspecting that the system is inimical to their interests."*

Folks, that is right out of the horses' mouths. That is the Rothschilds' view of central banks and they ought to know, as their family started the whole idea in England. These kind of people (CFR/TLC) still think that we are too stupid to catch on (apparently we are) and that they will end up owning everything, just as Jefferson predicted. If you still don't think that they think we are imbeciles, here is a confirmation from Lionel Rothschild:

> *"Can anything be more absurd than that a nation should apply to an individual* (private banks-RT) *to maintain its credit and with its credit, its EXISTENCE as a state, and its comfort as a people?"*

No, Mister Rothschild, nothing is more absurd, yet that is exactly what we have been doing since 1913. President Garfield addressed the problem of central banks when he said:

> *"Whoever controls the volume of money* (Federal Reserve Banks-RT) *in any country is absolute master of all industry and commerce."*

Here is more from Woodrow Wilson:

> *"A great industrial nation is controlled by its system of credit. Our system of credit is concentrated. The growth of the nation, therefore, and all of our activities are in the hands of a few men... We have come to be one of the worse ruled, one of the most completely controlled and dominated governments in the civilized world – no longer a government by free opinion, no longer a government by conviction and the vote of the majority, but a government by the opinion and duress of small groups of dominant men."*

He was talking about the founders of the CFR. Things have gotten much worse since his day. Teddy Roosevelt was very aware of what was going on, which is why he fought the monopolistic corporations so hard that he came to be known as the "Trust Buster".

During the 1920s, many farmers lost their farms as a consequence of exactly this kind of money manipulation. The manipulation by central banks of the gold inflow into the U.S. and then the outflow from this country was the primary cause of the Depression. When Roosevelt illegally confiscated most of the gold owned by Americans, exempting his friends' wealth through some clever loopholes, and then raised the official price of gold, the central banks and financiers made tremendous profits at our expense.

In the 1930s, the author of the Federal Reserve Act described how the act was supposed to stabilize the currency and help prevent recessions and panics. That was the excuse for the act. He went on to describe how the bankers had managed to change just a word here and there to totally redirect the functioning and results of the law. Most important was the changing of just one word in the phrase about maintaining the value of the dollar. The word was changed from *"must"* to *"may"*. That change paved the way for the bankers to manipulate the money supply for their own advantage, just as Rothschild had bragged, and led directly to the depression.

The Federal Reserve System's fiat (phony) money is backed by nothing. It has no value except people's faith that they will be able to use it to buy things. That faith is about to collapse among the foreign holders of trillions of excess dollars created out of thin air by the privately owned federal reserve banks for their own stockholders who are the large commercial international banks. When the economy does collapse, don't listen to the government's excuses and don't accept their "solutions", particularly the unconstitutional emergency powers they will try to use. It is the government's reckless spending and massive borrowing that has allowed the banks to create all that money out of thin air. It is also mostly the central banks around the world that we owe all that debt to.

"Governments sometimes do the right thing, but only after they have exhausted all the alternatives." - (Anonymous)

HERE COMES THE IRS

The second disaster that occurred in 1913 was the passage of the Sixteenth Amendment which allowed for direct taxation of individuals and paved the way for the income tax. There were all kinds of promises made about what would and wouldn't be done with the income tax in order to get it passed. Most of these promises have been broken. The people never did get to vote on the proposition. The state legislatures were promised money from Washington if they would pass the amendment. Even so, the amendment never got the approval of two-thirds of the states necessary to ratify it. Ohio had ratified it, but as more information became available, their legislature rescinded its ratification.

Later, when "enough" states had voted to ratify it, including Ohio, the government declared it ratified, even thought it did not have the needed votes. The income tax amendment never legally passed. It is invalid.

The graduated income tax was the second plank of Karl Marx's Communist Manifesto. Is it any wonder that it is the primary instrument of class warfare in the U.S.? The income tax opened the door to unlimited spending by the government, along with unlimited borrowing. With the ability of the federal government to confiscate everyone's income came unlimited credit. Now we find ourselves hopelessly in debt beyond anything we can ever pay back.

The founding fathers provided the prohibition against direct taxation of individuals for several very good reasons. Most of the things they feared, and some they didn't even think of, have come to pass after the Sixteenth Amendment overturned their prohibition. One of the reasons they prohibited direct taxes was the essential invasion of privacy that had to occur in direct taxation. Any questions about that one? If you think the current invasion of your privacy is bad, you haven't seen anything yet. One way or another, the feds are going to have complete records of everything you do or spend. Whether it is the IRS ID card, the health ID card, or the post office ID card, you will eventually carry a card with your entire history on it unless we elect a government committed to the Constitution.

As long as the Sixteenth Amendment stands, the government's hunger for money will continue to grow. Your pension funds are the next target, and bills have already been introduced in Congress several times to confiscate part of them. The only way to curb the government's thirst for power is to drastically limit its income to that necessary to carry out only the essential functions of government allowed by the Constitution. As long as the government has access to nearly unlimited money, the corruption, greed, and lust for power among politicians and bureaucrats will be impossible to control.

While we were both fighting and also financing Communism, setting up a socialistic welfare state, and sustaining an all-out attack on commonly held beliefs and moral values, the value of the dollar went down the tubes. Government, corporate, and personal debt exploded. Budget and trade deficits went out of sight.

The average chief executive in American corporations makes about 150 times the pay of the average factory worker. In 1994, among the corporations that eliminated the most jobs, the CEOs received an average pay raise of thirty percent. The taxes paid by the large corporations dropped from over twenty percent in 1960 to less than ten percent now. Now do you know why all those PACs give so much money to politicians in both parties?

While this was happening, the average real take-home pay of most workers in constant dollars was falling. They were working more hours and had less time for their families. Many families had to have two wage earners in order to buy a home. Some of the politicians who talk the most about helping the poor, receive huge contributions from huge corporations and sponsor legislation to further socialize the country.

FOREIGN POLICY

Our foreign policy is largely controlled by the CFR regardless of which party is in power. Many of the Secretaries of State from both parties have been members of the CFR and/or the TLC. Policy made at the CFR and TLC meetings is implemented by our State Department. It is against the law for private individuals to be engaged in conducting foreign policy. The CFR couldn't care less about the law. In fact, many of the laws that Congress passes are written by them to further their goal of a one-world government.

The question naturally arises as to why the CFR/TLC crowd would be so determined to have a one-world government that could so easily turn on them? For one thing, they believe that whoever controls the wealth of the world controls the governments of the world. In that, they are largely correct as Rothschild stated. However, they may be in for a very rude shock. The main reason that they want a one-world government is the concept of *Pax Romana.* A single world government would prevent wars, facilitate trade, control markets, eliminate competition, and allocate "jobs" to the cheapest labor sources in order to maximize profits. The underdeveloped third-world countries would supply the raw materials. The second-tier countries would supply the cheap but well-trained labor. The top-level countries would supply the technology, research, financing, and primary markets. Sitting atop the pyramid would be the CFR controlling the world, through the U.N. armed forces and their banks and corporations.

The CFR and their many satellite organizations have a real interest in world hunger and population control, especially in third-world countries. Look where so much of the money from their establishment foundations goes. That's why so much U.S. taxpayer money goes towards these areas as well. It sounds altruistic, but they really don't care how they accomplish either goal. This interest has nothing to do with humanitarianism. It has everything to do with control. Masses of hungry people start revolutions which is a no-no for efficient commerce or exploitation. If the multinational corporations can reduce the third-world populations and provide enough food for them, they will have enough people getting enough food and sustenance to produce the raw materials that they need. Keep them just happy enough that they don't want to lose what they have, and all will be well.

There is a second reason why the CFR/TLC is interested in improving world agriculture and food supplies. As countries turn away from more primitive farming methods, guess who gets to supply the farming equipment, chemicals, and seeds they need? The large multinational corporations. Guess who gets to finance all of this? The large multinational banks. Guess who ends up with the land if they can't pay back the loans? Guess who gets to sell them all of the "toys" people will want to buy if they do raise their standard of living? Get the picture?

The CFR/TLC and fellow travelers don't want to run the world for their personal satisfaction. That's for the elected officials whom they pretty much control. They want others to actually run the world while they dictate policy to them, so that they can make even more money. Like fire, which can be useful or deadly, the ambition for money, power and influence is a useful source of motivation when kept under control. When it gets out of control and becomes the reason for living, as in the cases of Hitler, Stalin, and so many others, the result is disaster. That is exactly where we are headed if the one-world superstate comes about--the greatest disaster and loss of life, not to mention freedom, that the world has ever known. The hundred million people sacrificed to Communism are a drop in the bucket compared to what will be necessary to bring about a single world-wide government. The one-world government proponents seem to forget that Rome started out as a republic and ended up as a dictatorship with a puppet senate. The same thing will inevitably happen to them, and us, with that kind of power available to whoever comes out on top.

FOREIGN POLICY

If you wonder what is really going on in the Balkans, there is a saying,

> *"Who rules East Europe commands the Heartland.*
> *Who rules the Heartland, commands the World-Island.*
> *Who rules the World-Island commands the world."* - H.J. MacKinder

> *"Russian generals have already smuggled 4,000 railroad cars of weapons and ammunition, including howitzers and antiaircraft missiles to the Serbs. Iran and other Islamic nations have smuggled plane loads of weapons to the Moslems."* - R. E. McMasters, Jr., *The Reaper,* January 25, 1995

See Chapter Seven for more information.

We may yet end up in a major war in Europe or Korea if Clinton needs a war to win re-election, or because of his roller-coaster foreign policy, whatever it is this week. The basic approach of our founders was to stay far away from entanglements with foreign nations. Unfortunately, we are now up to our ears in them. America cannot be the world's policeman. We can encourage and assist those who are fighting against foreign aggression or for their freedom and fundamental rights. Since the State Department is filled with mostly far-left liberals and CFR types, we need to clean house there. Then maybe we will start to negotiate treaties, when necessary, where we come out with some advantages instead of continuously giving away the farm.

Speaking of Clinton's foreign policy, if he has one, apparently he doesn't learn from history or doesn't care how dangerous the stand is that he is taking with Japan. The Japanese buy most of our debt. Without them, interest rates would go through the roof. When President Reagan began talking tough about the Japanese trade policies, they decided not to show up at a Treasury auction of bonds. As the time wore on, Treasury officials grew more and more nervous. Nobody was bidding on the bonds. Everybody else was waiting for the Japanese to show up. If the government couldn't sell its bonds, or the interest rate went way up to get rid of them, it would have been a disaster for the dollar and our economy. Finally, in the last few minutes, the Japanese bids came in. They had made their point. Reagan never said another word about Japanese trade policies.

Clinton's recent attacks and pressure to open their markets have deeply angered the Japanese. They have been forced to back down from the brink. They have lost face. Actually, Clinton accomplished nothing. The Japanese agreed to open some more dealerships for American cars. Big deal! The Japanese don't buy our cars for several reasons. One is quality. Another is national pride. One of the most obvious reasons they don't buy our cars is that Detroit is too stubborn to put the steering wheel on the right side of the car, except for Jeep. When they put the steering when on the right side, sales soared since the Japanese drive on the left side of the road, as in England. Clinton knows full well that opening a few more dealerships is not going to convince people to buy cars that are awkward and difficult to drive. The whole thing was just grandstanding for political gain at home. In doing this, he risked bringing our very fragile economy crashing down.

Right now the Japanese are between a rock and a hard place. They must inflate their economy to prevent their severe recession from turning into a full blown depression. They are buying shiploads of dollars with yen that their banks are creating out of thin air to try to reduce the price disadvantage the high value of the yen has produced for their products. The debate between the politicians and the bankers in Japan is growing very heated. The politicians want to pull the plug on the U.S. economy. The bankers are supporting the U.S. dollar, trying to increase exports to the U.S. to save their own skins. If the politicians win the argument, you had better have some substantial savings put away.

MEDIA MALFEASANCE

> *"Freedom House ranks the United States ninth in press freedom. Belgium, Australia, Switzerland, and Germany were the top-ranking nations in the 186-country survey. Many stories on the Clintons that are spiked (censored) by U.S. newspapers appear overseas, especially in the Sunday Times of London. The clout of large media chains and their pandering for advertising dollars are among the reasons for American press censorship. The steady growth of pro-freedom newsletters (such as MIA) is another sign of public dissatisfaction with the mainstream media."*- Don McAlvany, *McAlvany Intelligence Digest,* September 1994

Few would dispute that the major media in this country are owned and controlled by the far-left liberals. Polls taken of the owners and various high level people in the media clearly indicate that they hold values far different from the average American, and most of them feel that not only is it permissible to use their power and influence to change America to their viewpoint but it is their duty. That's why we have so little objective reporting any more.

Politics, particularly Presidential politics, has become a media circus. Image building and thirty second sound bites have become dominant factors in elections. Issues and character have become far less of a factor. Apparently, character and morality mean nothing anymore as long as you can project sincerity on the tube and tell people what they want to hear. Unfortunately, the media's influence on our culture goes far beyond politics. Violence, drugs, sex, infidelity, pornography, lying, and many other things that were once called vices are portrayed by the media as common, even acceptable behavior. While money may provide much of the media's motivation for pouring out all of this garbage on us, many of the leaders in the industry freely admit that they are deliberately using their power and positions to change society, and they are changing it. Just look at the crime statistics, the breakup of families, and the changing attitudes about morals in general.

Under our Constitution, the federal government has little to say about what goes on in the media, and this is as it should be. Government cannot solve this or most other problems. As the Constitution says, the power is left to the people. Until people stop watching the shows, reading the magazines, and buying the products that sponsor this stuff, it will continue. It is not up to the government to deal with these problems; it is up to you and I. The message goes where the money is. It's that simple.

EDUCATION

Evolution and its derivatives, existentialism, humanism, and liberalism (now neo-liberalism), first captured the colleges, where the future teachers were trained. The NEA became one of the most powerful lobbies in the country at all levels of government. Following the pattern established by Thomas Dewey, public school curricula began indoctrinating our children in these new world views. Now we have added one-world government, sex encouragement and radical environmentalism to the mix. Educators know very well that they are indoctrinating the leaders of tomorrow. We pour billions of dollars into this indoctrination program every year and turn out far too many people who can't even read or write.

Can the federal government help solve some of these problems? Yes, but not as a primary player. First of all, we have to restore government to its proper role and establish some measure of integrity in it. Until we as a people refuse to elect anyone without the character, integrity, honesty and principles that matter, we will get and deserve to get this kind of government that we are complaining about. The government can stop its unconstitutional funding of the liberal's educational agenda with your tax money. It can actually enforce laws already on the books in areas like pornography. It can stop funding sex encouragement classes in the schools. Ultimately, however, the people we elect to office should reflect our own values. Until the values of our society, which are heavily influenced by the media, return to some kind of commonly accepted moral basis, our downward spiral will continue.

In reality, the only thing that will seriously reverse the current trend is spiritual/moral revival on a broad scale, coupled with a tremendous amount of work on the part of many people. Evolution, humanism, and atheism have wreaked havoc on planet Earth in this century, and their influence and results will not be changed overnight or without a tremendous effort on the part of a large segment of our people. Are you willing to put forth that effort, or do you prefer that your children and grandchildren live in an even worse situation than what we face now?

Chapter Nine

The Rise and Fall of Evolution

"Darwin's descent has not a single fact to confirm it in all of nature." "Evolution is impossible to prove or disprove through science. We believe it because the alternative is unacceptable." - Sir Arthur Keith, British evolutionist

"It is when a people forget God that tyrants forge their chains." - Patrick Henry

This chapter is a continuation of Chapter Eight on how we got here. Since evolution is the root cause of many of our problems and needs to be dealt with at length, I have put it in a separate chapter.

In 1859, Charles Darwin published the first of his two famous books, *The Origin of Species*. This was not a scientific book. It was a book of fantasy or speculation. In the book, he used the phrase "let us suppose" 880 times. It was pure speculation, without a shred of solid scientific evidence. Darwin claimed that he got the ideas for this book when he visited the Galapagos Islands and observed the wildlife, particularly the finches. We now know that this claim was not true. His writings were, in fact, based on the rather voluminous writings of his grandfather, Erasmus Darwin. Much of *The Origin of Species* was obviously copied from his grandfather's work. Darwin never once mentioned his grandfather, much less gave him any credit. He insisted that the ideas expressed in his books were his own.

Darwin had no real scientific education. He had a divinity degree, but he couldn't succeed as a preacher. He inherited an interest in biology from his grandfather and eventually found himself on a ship going around the tip of South America. His grandfather had planted in him the seeds of doubt about God and the possibility of a natural explanation for life that might eliminate the need for a Creator. Darwin decided that maybe he could make a name for himself in championing his grandfather's ideas, and spent his time on that famous voyage looking for examples that might support these ideas.

Darwin's version of evolution (the basic ideas had been discussed and published for about 2000 years) was conceived in a desire for self-promotion, born in deception and brought to a position of prominence by the desire of his promoters to find excuses for their lifestyle, as will be demonstrated a little later. Evolution was basically a philosophy or a religion, searching for credibility by masquerading as science. The structure that has been built on the foundation provided by Darwin has, likewise, been riddled with speculation, fraud, and deception, as well as self-promotion. The most obvious examples of speculation and deception are in the field of anthropology, and in particular, the cavemen or "missing links" that my generation grew up with.

The Piltdown "man" was a deliberate fraud. His discovery was hailed with great fanfare in the scientific and popular press. Over five hundred doctoral dissertations were written about him, "verifying" his place in our ancestry before the fraud was uncovered. Were you ever told that he was a fraud? Of course not. The Nebraska "man" consisted of a tooth, from which an entire humanoid being was concocted. Raging egotism knows no bounds in creativity. The tooth was later proven to belong to a very modern pig. Java "man" consisted of a skull cap, three teeth and a piece of thigh bone. These fragments were scattered over seventy feet apart and found over a period of a year. There was no evidence that they even belonged together. He turned out to be a gibbon, and the discoverer eventually admitted that he had withheld evidence of this fact. Yet, he is still listed in current encyclopedias and textbooks as valid. Peking "man" turned out to be an ape, actually many of them. Yet, he too is still listed as one of the Homo Erectus series in encyclopedias and textbooks. Richard Leakey has discovered human remains in layers of rock that are, according to the evolutionary column, much older than any of these so-called missing links, which means that they weren't links at all. Yet, we are still being taught that they were. Ramapithecus, which was nothing more that a handful of teeth and jaw fragments, turned out to be an ape. Australopithecus also was eventually determined to be an extinct ape.

The Neanderthal man was indeed fully human. He was promoted as a missing link between apes and humans because he walked bent over, halfway between an ape's gait and human walking. Later analysis proved that this particular individual walked bent over because he suffered from severe arthritis and rickets. Before he died, his discoverer admitted that he had actually found three complete skeletons of Neanderthal Man. Two were perfectly normal people, and one had arthritis and rickets. He knew his discovery would only be of mild interest if the normal skeletons were revealed, so he hid the existence of the two normal skeletons and published his findings only on the crippled one, presenting it as an example of and evidence for the theory of evolution. It was immediately accepted, virtually without question, as was most of the "evidence" for evolution in those days, and he made his career out of that deception.

While Leakey remains an evolutionist, his discoveries have destroyed virtually everything we have been taught about man's "ancestors" for nearly a hundred years and exposed the desperation, or desire for status, that drove the hunt for these non-existent missing links.

Darwinism was based on uniformitarianism – the belief in gradual natural genetic changes – and natural selection of the "better" mutations. There were three primary problems with this theory. Uniformitarianism was never supported by the available evidence, and has now been abandoned. Natural genetic changes could never happen often enough to provide the needed genetic changes. This part of Darwin's argument was abandoned in the 1930s, along with most of his other suggested evidence, and replaced with the argument that radiation was the cause of most genetic changes, in what came to be called neo-Darwinism. Natural selection has now been shown to direct genetic changes back to the median, *not* enhance the oddballs. It effectively eliminates them by burying the damaged genes (recessive) or correcting them. Darwin was wrong on all three legs of the stool that evolution rested on, as most evolutionists now admit.

These three pillars of evolution have all been destroyed. Evolution, however, is still the accepted basic "scientific" theory of origins today and the foundation of our official state religion – humanism – which is the only religion allowed to be taught in our public schools. How did this come about?

Darwin's theories were rejected and scorned by most of the established scientists of his day. However, some of the younger scientists, desiring position and prestige, jumped aboard Darwin's boat and began promoting his theories. Probably the most famous of them was Thomas Huxley. He quickly became known as "Darwin's bulldog" and was largely responsible for the eventual acceptance of Darwin's suppositions in academia. Huxley powerfully promoted Darwin's theory in public but strongly disagreed with him in private. In private correspondence with Darwin, he repeatedly warned Darwin that his insistence on thousands of minute changes between species was completely unsupported by the fossil record and would be the undoing of the theory. His assessment was accurate, and his "prophecy" has finally come true in our day.

Even with the state of scientific knowledge in Darwin's day, it was recognized that the probability of even the smallest necessary transition step in generating a new species was far beyond the realm of possibility. When Huxley was asked how he could support such a theory despite the overwhelming mathematical probability against it, he acknowledged the paradox, and provided a enlightening answer. He said, "I believe in evolution, because I cannot bring myself to accept the alternative." In other words, forget the scientific evidence and the mathematical impossibility. His decision was based on personal preference and a desire to avoid personal responsibility for his actions. Does that sound similar to today's idea that everyone is a victim and not really responsible for *their* own actions?

Why was evolution Huxley's preference over creationism? Julian Huxley, the most recent in the line of Huxley evolutionists, said that the early supporters of evolution chose to champion it because "the idea of God interfered with their sexual mores." In other words, if there was no God, there were no standards of morality or behavior and they could indulge in sexual immorality with no pangs of

The Rise and Fall of Evolution - 199

conscience or sense of responsibility. Bertrand Russell, who was famous for seducing anyone he could, including minors, said of evolution that getting rid of God had freed him up from any moral restraints. Supporting evolution was a religious, or if you prefer, a philosophical decision that had nothing to do with any hard evidence. True scientific evidence was non-existent. Let me give you two quotes from top evolutionists.

> *"The descent of species is without a shred of actual proof."* - Charles Darwin

> *"Darwin's descent has not a single fact to confirm it in all of nature." "Evolution is impossible to prove or disprove through science. We believe it because the alternative* (Creator-RT) *is unacceptable."* - Sir Arthur Keith

I would bet that you were never given these quotes in school.

In 1959, a worldwide conference of top scientists from evolutionary fields was held in Chicago to celebrate the 100th anniversary of Darwin's publication of *Origin of Species*. One of the leaders of the conference was the head of the British Museum of Natural History. He had come to the conclusion that there wasn't a single scientific fact that could be demonstrated which supported evolution. During the conference, he stood up and asked if anyone could name one single thing that they knew for sure about evolution. There was a stunned silence, which quickly became a very embarrassing silence. Finally, one man stood up and said, "The one thing that I know for sure about evolution is that it should not be taught in our public schools." Probably the reason he made this statement is that virtually everything that is now taught in our schools is known to be false by the evolutionists themselves. It is too bad that our Supreme Court justices were not at the meeting.

Speaking of education, here is a quote from Stephen Jay Gould, today's leading spokesman for evolution. "You are no more important than a dry twig that you pick up in your back yard." Think about the ramifications of that statement. This is what the evolutionists have been saying to our kids for three generations.

If we are nothing but accidents with no purpose, as they contend, and if there are no standards of morality, is it any wonder that kids shoot each other when someone makes them angry? Why not? There are no standards. Everyone can make his own rules (situation ethics) and do whatever makes him feel good. The assault on the self-esteem of our children by their peers is bad enough but it is reinforced by teaching them that they are of no more value than a dry twig. It is no wonder that we are having an explosion of teenage drug use, sex, pregnancy, gangs, robbery, and murder. The bitter fruit of this philosophy is coming forth.

Evolution is not science, as most honest evolutionary scientists will admit. It is pure speculation, which doesn't work anymore. Thirty-six years after the admission in Chicago (that I just described), this false information is still filling our encyclopedias and textbooks. Karl Popper, a prominent evolutionist has said, "I have come to the conclusion that Darwinism is not a testable scientific theory, but a metaphysical research programme..." That's metaphysical as in speculative philosophy. Another prominent evolutionist has said that evolution doesn't even qualify as a scientific *theory*.

The two most proven and accepted *laws* of science are the First and Second Laws of Thermodynamics. In simplistic form, the First Law says that the total energy in the universe is constant. It never changes. The Second Law says that in *every* process, some of that energy becomes unavailable. Both laws are death blows to evolution. In essence, the first law says that at birth (creation), the universe started out with a given amount of energy that cannot change. Here is a question, Mr. Evolutionist: where did the initial energy come from? The Second Law says that everything is running down and that eventually all energy will be in an unusable form. Everything will be at the same temperature and no energy can flow from a higher state to a lower state. The universe itself will be dead. No processes of any kind will be taking place. Since the universe is running down but is not yet dead, it cannot always have existed, and therefore had to have a beginning in time and space. Here is another question, Mr. Evolutionist: what (who) existed before the physical universe, what (who) caused it to come into existence, and what (who) supplied the initial energy?

The Second Law applies not just to heat energy, but also to things like information transmission. Random errors creep into the process, even in the digital world, as any computer expert will tell you. In the analog world, signals degrade because of noise that cannot be eliminated. If you don't think that things are continuously degrading, take a good look in the mirror if you are over forty.

If you supply some controlled energy and creative intelligence to a bar of steel, a pile of sand, and some rubber, you can "create" a car. However, some of the available usable energy is lost in the process. More importantly, for the point that I am making, if you stop adding energy intelligently and let nature take its course, you end up with a pile of rust and very hard sand which is now useless. That is exactly what is happening to the whole universe. So, we are back to the question: *who* (and it had to be a who) supplied the initial energy and the *intelligence* to set up the physical laws and the design?

Evolution requires a process that is the exact opposite of the Second Law. In evolution, everything must become more organized and more complex as time goes by, all by accident and random events with no intelligent design or guidance. The evolutionists try to get around the problem with respect to the Earth by saying the sun supplies the energy to fuel the process. If you analyze their argument, it turns out to be pure sophistry. It doesn't wash. The process also needs intelligence. Evolutionists tell us, in effect, that the "intelligence" was supplied by random events and "natural" selection. Read that statement again. Does it make any sense to you? It sounds like an oxymoron to me.

In Darwin's time it was generally believed that the universe was infinite and stable. Then it was discovered that the universe was finite and expanding rather rapidly. With that discovery, the ship of evolution hit an iceberg and should have sunk. However, the cosmologists have been bailing like mad to try to keep it afloat. Einstein realized the creation/Creator implications of an expanding universe in his mathematics, and has tried desperately to doctor up his equations to come up with a static universe by "inventing" terms that actually didn't exist. It didn't work. His original (and accurate) equations of general relativity required that all energy, matter, space, and *time* grow from a single point of origin. Since this required extreme temperatures, it came to be known as the big bang theory.

To try to get around the implications of a single point of origin, the astronomers came up with the cosmic hesitation theory. This didn't solve the finite time/creation problem, so they then came up with the *infinite* hesitation theory, which is an oxymoron. When that fell apart, they tried the steady-state theory, which required spontaneous creation of energy out of nothing to fill the voids left by an expanding universe. Try to explain *that* to a thirteen year old. This violated both the First and Second Law of thermodynamics and was discarded, although one or two astronomers are still trying desperately to somehow make it work.

Next came the oscillating universe theory, in which the universe expanded for a while, then gravity overcame the initial energy outburst and pulled everything together again into one gigantic mass which caused it to explode again. It continually amazes me what some people will do or continue to believe, despite all the evidence to the contrary, in order to avoid dealing with God as Creator and the implications of His existence. This theory also violates the Second Law and leaves the question of where the universe originally came from unanswered. Worse yet, accurate measurements have shown that there is not even one-tenth enough matter in the universe to pull it back together. Scratch that one.

Here is the essence of the problem of the expanding universe for the atheist. By measuring the expansion rate, you can mathematically run the tape backwards. By doing so, you come up with a figure for the age of the universe: about sixteen billion years, as a maximum, which agrees with numerous other kinds of measurements. Before that, *what*? From the mathematics of quantum mechanics, you can go back to the time when the universe was just 10^{-43} seconds old. Before that, everything falls apart.

Since both time and space depend upon the existence of energy and matter, time as we know it did not exist before the universe was created. The question is, "what or who *did* exist?" because *"nothing"* cannot somehow create *"something"* out of *"nothing"*. That is the problem of the First Cause, which evolutionists have never have been able to deal with. You cannot have physical laws without a lawgiver (definer). You cannot have a finite amount of energy that never varies (First Law) without that energy having an original source. Therefore, you have to have a source of energy

The Rise and Fall of Evolution - 203

and intelligence that transcends space and time, i.e., GOD. Robert Jastrow, a prominent astronomer, noted at the end of his book, *God and the Astronomer*, that after the scientists had struggled up to the top of the mountain of knowledge (of origins), that when they finally pulled themselves up over the last rock, they found a group of theologians had been there all the time.

The unescapable implications of all of this, which are much more complex than I have described, led Fred Hoyle, a famous astronomer and mathematician, to reverse his stance as an atheist and a believer in evolution. Both physics and mathematics left him with no other choice but to acknowledge the necessity of a Creator. From his book, *Evolution from Space*, with co-author Chandra Wickramasinghe, he states:

> *"Once we see, however, that the probability of life originating at random is so utterly minuscule as to make it absurd, it becomes sensible to think that the favorable properties of physics, on which life depends, are in every respect, deliberate.... It is, therefore, almost inevitable that our own measure of intelligence must reflect higher intelligences...even to the limit of God."*

In the magazine *Nature* (Nov. 12, 1981), Hoyle stated: "The chance that higher life forms might have emerged in this way (evolution-RT) is comparable with the chance that a tornado sweeping through a junk-yard might assemble a Boeing 747 from the materials therein." I hesitate to disagree with so famous a mathematician, but it seems to me that it would be more like a billion tornadoes roaring through a billion *different* junk yards and building a billion *identical* 747s. At any rate, this is only an illustration that attempts to conceptualize the utter impossibility of evolution actually occurring even if conditions were ideal, which they were not. Regardless of the number of 747s spontaneously constructed, you know full well that it could *never* happen. Yet the probability of spontaneous generation of life and subsequent evolution is so far beyond this example that it is incomprehensible, as we shall see a little later.

The number of bits of information stored in the human genetic material of one person is greater than all the words in every book ever written. Yet, the evolutionists tell us that we should believe their fantasy which is the same as saying that something came along and randomly scrambled the letters in some of the words in many of those books and *presto*, whole new books appeared that made sense. Just because this was supposed to have happened in living organisms doesn't make it any easier. In fact, genetic changes are much more difficult because of the genetic repair mechanisms, redundant code, and ability to suppress bad code that living organisms possess.

One of the things you might not have picked up in Hoyle's illustration is that tornadoes are destructive, not constructive. They blow things apart; they do not assemble them. Is his illustration therefore invalid? Not at all, because the various forces of nature and physics, such as the Second Law of Thermodynamics, are likewise destructive. Radiation does cause mutations, but these mutations are always *destructive*. Radiation destroys the encoded information in the genetic material. The idea that background radiation could cause billions of gradual changes that were positive, and in the right sequence at the right time for just the right circumstances in order to provide millions of viable species is incomprehensively absurd. In over 3,000 successive generations of fruit flies in which mutations were deliberately induced at a high rate with radiation and very carefully selected, not one genetic improvement was shown. We are asked to believe that the combination of random destruction of the genetic code and natural selection, which the evolutionists can't even agree on or adequately define, could somehow very accurately do millions of times what we cannot do one time, with the best equipment and all of our intelligence, dedication, and knowledge. But, you ask, doesn't the fossil record record it happening? Absolutely not. In fact, just the opposite is the case, as we shall see a little later.

Speaking of genetic material, the geneticists have discovered that all of humanity descended from one female. Guess what they named her? Eve, of course. That fact alone completely destroys evolution. If it isn't obvious why, think about it for awhile.

According to Hoyle's calculations, the chance of life originating from a random process is one in $10^{40,000}$. We will take a look at that number a little later. This also assumes that conditions were perfect for the formation of life. They weren't. Darwin himself was aware of the problem, even on a very primitive level, without the current knowledge of genetics and microbiology, and tried to get around it by imagining (fantasy again) a primordial soup with all the right ingredients and the required reducing atmosphere (no oxygen) essential for such spontaneous generation. For about a hundred years, evolutionists *assumed* and stated as fact (as they have done with many other fantasies) that the early Earth had a reducing atmosphere. Now we know that it *did not*. There was oxygen present long before life could have spontaneously generated. Therefore, spontaneous generation of life *could not have happened.*

Just to give you a taste of the problem, if the Earth has oxygen, then the chemicals required to start life would have been removed by oxidation. Furthermore, amino acids cannot join together in the presence of oxygen. However, if there was no oxygen, there would have been no ozone and the ultraviolet radiation would have quickly destroyed anything resembling life. We now know that there has always been oxygen on the planet, and certainly long before anything like life was here. Beyond that, the chemical processes needed to form proteins move opposite the direction required by evolution, and the available energy sources for the necessary energy would have destroyed the protein products much faster than they could have been formed.

It gets even worse. Amino acids come in two forms, right-handed and left-handed, in roughly equal amounts. They are chemically equivalent. There is no known natural process that can separate the two forms. Yet, all living things, from man to viruses, consist only of left-handed ones. How did "nature" select only left-handed ones? No one has the slightest clue. Sugars demonstrate a similar problem. Nature makes both right- and left-handed forms, but living things consist only of right-handed forms. As Yul Brynner, while playing the King of Siam, used to say, *"'Tis a puzzlement."*

Let's go back to the fossil record. Doesn't it demonstrate evolution? Absolutely not. Huxley repeatedly attacked Darwin's insistence on gradual, minute changes because the fossils showed no such thing. Quite the opposite. Life in myriad and complex forms suddenly appeared in Cambion layer rocks. Other forms suddenly appeared, fully developed, in other layers. This sudden appearance of new life forms in different layers was hard evidence against Darwin's gradual change theory, as Huxley pointed out. Darwin's answer was that the reason for the apparent contradiction was that we had not dug up enough fossils yet, and that someday the missing links would be found. Well, we now have mountains of fossils, and guess what, not one missing link. If evolution actually happened, we would have millions of examples of missing links.

Evolutionists have tried to make missing links out of a number of fossils, and they have all been proven false. Perhaps their best candidate was a "pre-bird" that was supposed to be a step between reptiles and birds. This fossil imprint was named Archaeopteryx. They claimed Archie was a missing link, because he had claws on his wings and he had teeth "left over" from his reptilian ancestors. Unfortunately for the evolutionists, a modern bird had been found that has similar claws to help him hold on to cliffs. In addition, other fossil birds that are truly birds have been found with teeth. Still worse, plain old birds have been found that predate Archie in the evolutionist's cockamamie dating system, so he couldn't be an intermediate step. He is now classified as a true bird. So it has been with all of the supposed missing links. There aren't any, because there never were any. Evolution did not happen.

Huxley was right. The lack of any transitional fossils doomed Darwin's gradual step evolution. Few evolutionists will try to defend it today. They admit that the fossil record shows only fully formed differentiated species. So they have replaced it with an even sillier idea to try to conform to the actual record. in 1940, Dr. Richard Goldsmith proposed the "Hopeful Monster" theory which basically said that somehow there was enough massive genetic change occasionally that a reptile could lay an egg, and out would pop a bird when the egg hatched.

Even if you haven't fallen off of your chair laughing, you can still appreciate that the scientific community responded to this idea with a great deal of well-deserved ridicule and derision. No possible mechanism could explain such a massive and accurate rewriting of the genetic code. And it had to be rewritten not once, but several times, as the codes are duplicated in the DNA. That's not all. This same accidental rewriting of the genetic code (roughly equivalent to *correctly* writing down every word in every book that has ever been published) had to happen twice identically and at virtually the same time, with one accident being male and one being female for the new bird to survive and procreate. And, of course, both of them would have had to be fertile after all of that genetic rewriting, which does not happen even when very close relations like donkeys and horses are interbred. The offspring are sterile. The boundaries of variation are rigidly fixed. Similar impossible dual accidents would also have had to occurred countless times in order to produce all of the millions of different species. As many evolutionists pointed out, this was about as ridiculous as you could get, and there wasn't a shred of evidence that such a thing had occurred or could occur and there was no clue as to any mechanism that could possibly cause it.

However, as it became even more evident that the fossil record had become evolution's biggest problem rather than its primary evidence, a strange thing happened. The evolutionists began embracing Goldsmith's hopeful monster. There really wasn't anywhere else to go, other than to throw out evolution entirely.

Evolution has proven to be the most elastic theory ever dreamed up. When the facts falsify it, the evolutionists construct an even more absurd explanation. One very prominent evolutionist has said, "Evolution is an adult fairy tale". Anything, no matter how absurd, is better than dealing with the implications of a Creator. L, T. More said: "The more one studies paleontology, the more certain one becomes that evolution is based on faith alone; exactly the same sort of faith which is necessary to have when one encounters the great mysteries of religion.... The only alternative is the doctrine of special creation."

I would disagree with Dr. More as to degree. Evolution has no solid scientific or historical evidence to support it. On the other hand, the life, death, and resurrection of Jesus Christ are by far the most thoroughly documented *facts* of ancient history. If you doubt that statement, read one of Josh McDowell's books, starting with *Evidence That Demands a Verdict*.

The First and Second Laws of thermodynamics demand a creator. The Theory of Relativity demands a Creator, and on and on. When we look at probability theory, it is obvious that believing in evolution, once the true facts are known, requires far more of a leap of faith into a dark chasm than anything required of religion.

Let's go back to the hopeful monster theory. No one was going to accept anything called the hopeful monster as serious science, so it had to be renamed and is now known as punctuated equilibrium. This sounds much better, doesn't it? However, the idea is still the same silly nonsense, despite the evolutionists' efforts to dress it up. For a hundred years, the very foundation of evolution was uniformitarianism. When that proved untenable, they threw out the foundation but tried desperately to keep the building intact. Needless to say, it is crumbling rapidly.

The evolutionists knew full well that the idea of a one-step lizard-to-bird jump would never fly (pun intended). So, Julian Huxley, among others, proposed that the jump occurred in *perhaps* four or five steps. This only made matters worse. There were still no examples in the fossil record of the millions of intermediate steps still required to make this scheme work. The necessary rewriting of the genetic code was scarcely less than in Goldsmith's hopeful monster scheme, and the probability of the accidents happening in the correct sequences for millions of changes soared far beyond the realm of possibility. The next time you hear or read the term "punctuated equilibrium", picture a bird hatching out of an alligator egg.

Many things have been discovered that blow away the assumptions of evolution. I will mention only a few. The amino acid sequences of Cytochrome C show that there is less difference between a single-celled bacterium and a horse than between a single-celled bacterium and a single-celled yeast. The lamprey is supposedly more primitive that a fish, but the Cytochrome C shows that a lamprey is more closely related to a human than to a fish. It is not just the Cytochrome C. Hemoglobin differences show the same kinds of problems.

The material in the preceding paragraph comes from a book by Dr. Michael Denton entitled *Evolution: A Theory in Crisis*. It contains much more, and destroys such arguments as homologous structures that evolution has heavily relied on. In his book *Homology an Unsolved Problem*, evolutionist Gavin de Beer also showed how homology had been falsified. The whole idea of the scientific method of investigation is to propose a reasonable theory and do everything you can to falsify it. If you can, the theory is scrapped. Evolution is the only supposedly scientific concept in which the whole approach has been reversed. It was a fantasy to start with, without a scrap of hard evidence to base the concepts on, and billions of dollars have been spent trying to find anything that could possibly be interpreted as supporting it, even if evolution had to be assumed to get that interpretation. Anything that looked like it could falsify evolution was simply ignored.

In Darwin's time, his books opened up a number of new fields of investigation. Eager young scientists rushed into these areas to make names for themselves. Evolutionists quickly captured the review process of scientific publications or started their own. Opposing views and evidence was shut out. Soon, young Ph.D. students could not get research projects approved that did not accept evolution. Open scientific investigation died in related fields. Possible falsifications were denied publication. Today, even avowed evolutionists are openly questioning the assumptions and the conclusions based on them and the whole house of cards is falling apart.

Do you remember the intriguing geological column pictured in your textbooks or encyclopedias? All those layers of rock were shown with different kinds of fossils arranged in ascending order and very scientific-sounding names. Do you know where the column came from? It came out of thin air, as did the time-table. It was pure speculation, dreamed up to try to give evolution a foundation. No such "column" exists anywhere on earth. The closest they can come is about three layers, and none of these have three consecutive layers in the sequence. There are always missing layers. Worse, in many cases the layers are in the *wrong* order.

Evolutionists try to explain these wrong orders by claiming they were folded over or layers slid over one another. In a few cases, this is obviously true. In most cases, this explanation is impossible. The geology and the required mechanics refute it. These are sedimentary rocks that were clearly laid down in the wrong order for evolution to be true. The fossils are in the wrong order, and were laid down in that order because the order that various fossils were laid down in has nothing to do with evolution, but with the timing of when the sediment swept over them and where it carried them.

Actually, fossilization can only take place with rapid burial. Otherwise the fossils disintegrate before becoming fossils. The slow deposition of silt that uniformitairism requires, and which the evolutionists claim formed the fossils, can never do so. There is no example of such a process now taking place. The fossils were caused by rapid burial of live victims, or soon after their deaths, by massive hydraulic action, as in a massive and almost certain world-wide flood of great depth and force. That's how the different layers can be laid down in different orders in different parts of the world. That's the only way some layers of certain kinds of fossils can be intermingled with other layers which don't come anywhere near each other in the fictitious geological column. That's the only way sea creatures can be intermingled with land animals as is frequently seen. The whole timetable the evolutionists thought up to (supposedly) give enough time for evolution to work is pure fantasy. Now we know that even that timespan is hopelessly too short to even start the process, even if there was such a process. They are stuck with their fictitious dating scheme, having taught it as fact for over a hundred years.

Here are just a few things that throw the whole idea of the established geological column into the ash can. How about man-made objects found in coal deposits that were supposedly laid down hundreds of millions of years before man appeared? How about petrified trees standing up through three different layers of the column, where each layer is supposed to have taken hundreds of millions of years to lay down and gradually creep up the tree while it stood there somehow, immune to all the usual forces of decay? Impossible! How about layers of rock deep in the grand canyon that date, by the evolutionists' methods, as far younger than the layers near the top? How about human skulls and skeletons found in rocks supposedly millions of years older than man? Then, there are the hoofprints of modern horses as well as human prints, found alongside of dinosaur foot prints in the Soviet Union. The list goes on and on.

This whole dating scheme is known as a tautology or circular reasoning. This is how it works. The rocks are dated by the fossils they contain. The fossils are assigned to certain dates by the rocks they are in. Get it? There is no reference point. It gets worse. The rocks are dated not by all the fossils they contain, but only by certain "index" fossils which leads to all sorts of problems when the index fossils say one thing and some of the other fossils indicate some other time. Contradictions are ignored. Evolution is full of such tautologies.

The Earth is a uniquely designed ecosystem, with thousands of characteristics that must be there for life to exist. Many of these must be within very narrow limits. Just a few of them include the size, mass, color, and luminosity of the (one) sun; the size, mass, and makeup (material) of the earth and its distance from the sun; the number, size, mass, and distance of its *one* satellite; the axial tilt, surface gravity, magnetic field, thickness of crust, and reflectivity; and the oxygen quantity, oxygen-to-nitrogen ratio, ozone levels, carbon dioxide levels, water vapor levels, and many other properties of the atmosphere.

Several people have calculated the probability of such a planet coming into existence by some unknown natural process. It is so far below any chance of even one such planet in all the billions of stars that the probability of such an event is clearly impossible. Yet, the earth is here. Therefore, say the evolutionists, it obviously happened. Yes, it happened, but *not by accident or natural processes.* It could not possibly have happened by accident. Some of the boundary conditions are so narrow and precise that they could not have happened by accident all by themselves, much less in combination with all the others.

In case you are wondering, none of the theories about how the solar system came into existence work. Some planets and satellites rotate in the wrong directions, revolve around the parent planet in the wrong direction, are in the wrong plane, or otherwise falsify any natural explanation anyone has come up with.

One of Darwin's hopeful arguments was that embryology recapitulates the stages of evolution. This was one of his strongest arguments, which, like most of his stuff, he borrowed from someone else. This ideal was thoroughly debunked and thrown out by the evolutionists themselves in the 1920s as were most of his ideas he put forth as possible evidence for evolution. Yet it is still found in textbooks today, over seventy years later, and is still used by evolutionists in debates with creation scientists, even though they know it is nonsense.

Speaking of creation scientists, their ranks are growing by leaps and bounds. A number of Russian scientists told Dr Duane Gish that they knew evolution was a bunch of nonsense all along, but that they were required to teach it anyway. In America, we are still required to teach it, but unfortunately, most of our teachers and professors haven't realized that it is nonsense or are afraid to say so as in several recent cases. Maybe the Soviets had a closer look at its consequences in terms of the inevitable results of following its philosophy in the political and social area. However, we are now reaping the rotten fruits of its teaching in the social and political arenas to a great extent, and we can't seem to see the connection. Social Darwinism has become the scourge of the Twentieth Century affecting virtually every aspect of life and providing the philosophical foundation for nearly all the "isms": wars, genocide, and much modern racism as well as the destruction of much of the moral and spiritual values that propelled this nation to the top of the ladder.

Let's look briefly at the area of probabilities. Some of you may not be familiar with scientific notation, so here is a brief review: $10^1=10$. $10^2=100$. $10^3=1000$. In other words, every time the exponent increases by just one, the resulting number increases *ten times*. By the time the exponent gets to just 10, the number becomes almost incomprehensible in terms of real things that we are familiar with. Do you remember that number of $10^{40,000}$ as the probability for life originating from any random natural process? Let's try to put that in some sort of perspective. There are trillions of electrons in each of your cells. There are trillions of cells in your body. There are billions of people on earth, not to mention the animals, plants, rock, oceans, crust, core – the whole earth. The earth is a pea compared to a basketball, in regard to the sun, which is a small star, one of billions in our galaxy, and there are billions of galaxies.

With all of that, there are about 10^{80} electrons in the whole universe. The probability of something happening of 1 in 10^{80} power is like telling you to take a pair of electron tweezers and pick one certain electron out of the whole universe. Where do you start looking for this electron? Is it in you? Is it on earth? Is it in this galaxy? Is it in some of the interstellar dust? If you are starting to get the idea of what the evolutionists are asking us to believe, hang on. We haven't even started yet. That's for 10^{80}. What about 10^{81}? That would require 10 universes. What about 10^{82}? That would require 100 universes. How many universes would there have to be to get to $10^{40,000}$? Forget it. We have no way of expressing it, except by scientific notation, and certainly no way of comprehending it. The evolutionists have no answer to the probability problem except that they believe it somehow happened, no matter how impossible. Hoyle's math is accurate, and his assumptions are quite reasonable. Their only answer is that somehow the real world doesn't apply to biology, and somehow evolution must have happened because the alternative is unacceptable philosophically. In other words, *blind* faith against all reason.

Speaking of blind, one of Darwin's biggest problems, by his own admission, is the human eye. As he said: "To suppose that the eye, with all its inimitable contrivances for adjusting the focus to different distances, for admitting different amounts of light, and for the correction of spherical and chromatic aberration, could have been formed by natural selection, seems, I freely confess, absurd in the highest possible degree.... The belief that an organ as perfect as the eye could have formed by natural selection is more that enough to stagger anyone." Yes, Mr. Darwin, that belief is truly absurd, and you haven't even scratched the surface. That was long before we knew just how complex the eye truly is. Neither he nor anyone else has come up with an explanation of how the eye could develop piecemeal under his minute change-and-selection process or by any other process. Everything has to be there all at once for it to provide the "advantage" necessary under his scheme. But the eye is so incredibly complex that it could not have happened with any possible evolutionary mechanism. Also, we now know that the "evolution" of the eye would have had to happen five different times, independently, on completely separate tracks for different species. That's impossible times impossible five times over. The evidence against evolution is pouring in from almost every related scientific field.

I can't leave this area of multiple designs or factors instantly coming into existence at the same time which evolution can't allow or explain, without mentioning my favorite beetle, and it is not a musical one. The Bombardier Beetle is a fascinating creature. He has a unique defense system. When threatened, he ejects a gaseous mixture that immediately explodes with considerable force in a predator's face. If you are wondering what the big deal is about that, I will be happy to tell you.

The beetle makes and stores a gas that is so extremely explosive, it would blow him apart the instant it was made if, at the same time, he didn't manufacture an inhibitor to mix in with it. Can you imagine the changes in the genetic code required to suddenly start making either one? In the evolutionary process, which one did he "learn" to make first? If the gas, boom! That's hardly an advantage. If the neutralizer, why? It offered no advantage for the natural selection process and would disappear as the natural processes in the genetic code buried it as an aberration. Also, why would it be manufactured in the first place? Stay with me, it gets even more interesting. How does he get the gas out? Well, there is a tube that ejects the gas. Which came first, the gas (boom), the tube (why), or the inhibitor (why)? None of the three do him any good, because the inhibitor keeps the gas from exploding, inside or out. Well, random mutations (according to the evolutionists) designed yet another chemical that neutralizes the inhibitor. But we still have a problem. If he adds the neutralizer inside the body, he blows up. There is no way he can add it outside the body. Now what? Well, random mutations went back to the drawing board and designed a very nice solution over a few million generations, where none of this worked and provided no advantage to keep this "evolution" going. The gas is expelled at just the right velocity and the neutralizer is added at just the right split-second so that the time it takes it to spread through the gas is precisely the time needed to allow the explosion to occur at just the right distance for maximum effectiveness. Great designers, those random mutations. I wonder how many beetles they blew up before they got it right. What utter nonsense.

According to Dr. Karl Popper, one of the most highly respected scientific philosophers in the world, this theory of origins does not even qualify as a scientific theory, much less a scientific fact. Popper is not alone. In 1980, the British Museum of Natural History said the same thing in an exhibit, as have many others. However, evolution is taught as fact in our schools under the guise of science, while the historically proven facts about Jesus Christ cannot even be mentioned and the Christian foundations of this country are erased from the public school history books.

I haven't even scratched the surface of the various areas of scientific evidence that are falsifying evolution. For instance, there is a tremendous amount of valid scientific evidence that the earth is far younger than the evolutionists claim. Just some of this evidence includes the salianation rate of the oceans, the thickness of ocean sediments, the thickness of the dust on the moon, the Earth's decaying magnetic field, population dynamics, and many others. Evolutionists don't even try to explain away this evidence, because they can't. To them, evolution is true by definition (assumption), therefore such evidences are only anomalies that somehow, someday, can be explained. In any other area of science, such "anomalies" are better know as falsification. So much for scientific integrity. There are a number of books listed in the Appendix that cover many more of these areas in much greater detail. If you have any doubts about evolution being a pile of discredited fantasies, I hope that you will read some of them.

If I may be allowed to quote a Bible verse, it is: "By their fruits you shall know them." The fruits of evolution have been unparalleled death, destruction, misery, captivity, and waste. Some step forward! If you wonder why this information is in this book, it is because evolution, as a basic life philosophy, is largely responsible in an amazing variety of ways for the mess we now find ourselves in. It will be extremely difficult to work our way out of it without dealing with it as the phony pseudo-science that it is.

As evolution fades from the scene as a substitute for a Creator, all kinds of religious/spiritual philosophies will and have come to the forefront. The idea that man can keep evolving until he achieves some kind of super status and is somehow (for those of you who read Calvin and Hobbes) transmogrified into some kind of god is pure fantasy. Egotism has no bounds. This lie (that we can be a god) is as old as mankind. It is represented in the second chapter of Genesis by the serpent's statement to Eve. It was deadly then, and it is just as deadly now.

There are large numbers of books and groups that purport to know how you can achieve this god-like status. These ideas are put forth by people who can't tell the difference between a Creator God, who can cause a whole universe to come into being from nothing, and a creature who can't manage his own emotions without a lot of outside help and training. I wouldn't waste my time reading or listening to the stuff. Most of it comes from the ancient mystery religions of Babylon and it doesn't make any more sense now than it did then. You and I are creatures, neither accidents nor gods; we had better come to terms with that before we die. Either the universe, Earth, and people are senseless random accidents *or* we were designed and created by Someone for a specific purpose.

I think it would be fitting to close this chapter with a quote from a famous astronomer, Percival Lowell.

"Surely, the undevout astronomer must be mad"

or, as the Bible puts it, "The heavens declare the glory of God".

Chapter Ten

Two Views of the Future

"Politics for the most part now means promising more and more to the people, while taking more power and wealth from them to concentrate power in the state." - R.J. Rushdoony

ONE-WORLD-GOVERNMENT PROPONENT'S VIEW

You don't need to read books that attack the CFR/TLC crowd to find things that should scare you. You can read their own writings which express their goals clearly. Apparently they don't think ordinary people will read their writings or can do anything about their agenda. Basically, they believe that only a one-world government can bring peace, stability, and a dependable, profitable world. By controlling the variables, namely people, there is no limit to the profits and power they can have.

The one-world-government proponents have two powerful forces working in their favor besides their money and influence. The first one is the drive of politicians to make ever more laws and expand their own power and the bureaucrats' desire to make ever more regulations and expand the bureaucracy. After all, aren't legislators elected to pass laws? Aren't bureaucrats hired to dream up and enforce regulations?

The second force working in their favor is the desire people have for security as well as someone to take care of them. This desire makes the collectivist/socialist politicians appealing to many, particularly to poor people. It has been estimated that the life expectancy of a new democracy is about two hundred years. It takes about that long for the poorer classes to learn that they could elect people who would vote them sustenance from the public treasury until the country was bankrupt. Then a dictator or foreign power would take over. We are about to prove that estimate correct.

Those who pooh-pooh the idea that the CFR/TLC crowd wants to take us to a one-world government must have their heads in the sand. The leaders of these interlocking groups have made their plans quite plain. Their goal is for the U.N. and a bunch of controlling bureaucracies to run the world and keep everything nice and tidy so that they can live high on the hog.

> *"U.S. President Bush and Soviet President Gorbachev arrived yesterday on this Mediterranean Island for a summit conference beginning today during which both hope to start the search for a NEW WORLD ORDER.."*
> - *New York Times* (12/1/1989)

The term "New World Order" is not some vague reference to some foggy idea. It has a very specific meaning: a world run by the U.N. with one super-powerful leader.

> *"There also exists another alliance – at first glance a strange one, a surprising one – but if you think about it, one which is well grounded and easy to understand. This is the alliance between our communist leaders and your capitalists."* - Alexander Soizhenitsyn

Actually, these two groups have been working together since before 1917. American industrialists financed the Communist Revolution, and have been rescuing the Russian economy ever since with considerable help from the American taxpayer.

> *"We at the executive level here were active in either the OSS, the State Department, or the European Economic Administration. During those times, and without exception, we operated under directives issued BY THE WHITE HOUSE. We are continuing to be guided by just such directives, the substance of which were to the effect that we should make EVERY EFFORT to so ALTER LIFE in the United States as to make possible a comfortable MERGER with the Soviet Union."* - H. Rowan Gaither, President of the Ford Foundation.

Two Views of the Future - 219

In order to merge the U.S. with the Soviet Union, these one-worlders must make us so dependent on the government, and so comfortable with socialism, that we will willing merge with them as we will be so much like them.

In reference to the group that became the CFR, Woodrow Wilson said: "There is a power so organized, so subtle, so watchful, so interlocked, so complete, so pervasive that prudent men had better not speak above their breath when they speak of it."

In case you don't think George Bush was totally behind this New World Order concept, despite his hundreds of positive references to it and his actions in support of it, here is one of those references.

> *"This is an historic moment. We have in the past year made great progress in ending the long era of the cold war. We have before us the opportunity to forge for ourselves and for FUTURE GENERATIONS a New World Order,* (note that New World Order is always capitalized-RT) *a world were the rule of law,* (U.N. law-RT) *not the rule of the jungle* (nationalism-RT) *GOVERNS the conduct of the nations. When we are successful, and we will be, we have a chance at the New World Order, an order in which a credible UNITED NATIONS can use its peacekeeping role* (U.N. armed forces-RT) *to fulfill the vision of the U. N. founders."* (Virtually all of whom were communists or socialists-RT*)* - George Bush.

> *"The ultimate goal of a supra-nationalist world community* (One-world government-RT) *will not come quickly...but it is not too early to prepare ourselves for this step BEYOND nation-state."* - Henry Kissinger.

National boundaries are to disappear in this New World Order. Many prominent people have spoken out against the goals of the CFR/TLC, but they have been ignored by the liberal press.

"In my view the Trilateral Commission represents a skillful, coordinated effort to seize control and consolidate the four centers of power–political, monetary, intellectual, and ecclesiastical. All this is to be done in the interest of creating a more peaceful, more productive world community. What the Trilaterals truly intend is the creation of a worldwide economic power superior to the political governments of nation-states involved. They believe the abundant materialism they propose to create will overwhelm existing differences. As managers and creators of the system, they will rule the future." - Barry Goldwater.

"The New World Order agenda has been pursued relentlessly since the end of World War Two (a critical part of which was to form the U.N.-RT) *with no interruptions in strategy and only occasional shifts in tactic. The final question we need to answer is, 'Is it really so bad?' My answer is yes. Yes, because in the process we will lose more of our freedoms and most of our wealth. As the insiders'* (CFR/TLC-RT) *age-old dream of a New World Order comes closer and closer to realization, our personal options will be narrowed. We will, as Carroll Quigley said in Tragedy and Hope, 'be numbered from birth' and 'be followed through life.'* (Hillary's big push-RT) *It is George Orwell's nightmarish vision of the future come true."* - Larry Abraham, *Inside Report*

"George Bush looks upon himself as the chairman of the board of Skull and Bones International. He gets together with all the leaders of the world, and they decide this, that, and the other."...."Some of these fellows have to wake up and realize it's not 1939 anymore. Most of these folks (the Bush crowd) are globalists. They believe

> *in subordinating American Sovereignty to some globalist New World Order. When I see European countries giving up their currencies, giving up their control of trade...When I see George Bush engaged in the unilateral economic disarmament of his own country, I say watch out."* - Patrick Buchanan.

This is not a recent agenda. Since their founding in 1776, the Illuminati have been deeply involved in bringing this one-world Socialist order to pass. They were very active in the French Revolution and were well known to our own government.

The income tax amendment and the Federal Reserve Act in 1913 were tremendous boosts to the plans of the bankers and socialists. Then, the central banks caused the Depression by first flooding the country with gold, increasing our highly leveraged money supply that went mostly into the stock market, and then withdrawing it, causing serious deflation and collapsing the stock market. That brought Roosevelt to power, and he dramatically advanced the one-world cause and started the whole long march to a powerful nanny government and Socialism. Among other things, Roosevelt confiscated the gold of most Americans, but exempted many of his friends by means of loopholes inserted in the law. When he raised the price of gold more than fifty percent, they made huge fortunes from the profits, while most of the rest of America was selling apples on street corners or doing whatever work they could find to do.

Lyndon Johnson decided to expand the safety net to everyone he could. Since then we have spent/wasted trillions of dollars on the welfare bureaucracy, there are more people below the poverty line than ever, and educational achievements have plummeted. We are on the verge of bankruptcy, and it will not take much to bring down the whole house of cards built on our fiat money system and fractional reserve banking.

The Communists, Socialists, CFR/TLC, new-agers, neo-liberals, multinational banks, and corporations have a far different version of the future for us than we do. It amounts to total control of every aspect of our lives (us peons) by the U.N. and the many peripheral bureaucracies they will set up, like NAFTA, GATT and the World Trade Organization. The following long quote is from *The Phoenix Letter* by Dr. Anthony Sutton in 1986.

> *"The real enemy of free enterprise societies is not world communism-it is the multinational corporation which gives technological and financial subsidy to non-viable socialist economies.*
>
> *"..Communism is not an economic threat. Socialism does not work, period. Socialism can only advance through injections of free enterprise technology. Lenin knew that.*
>
> *"The danger is from financial monopoly capitalism-the corporate empires built by Rockefeller, Morgan, Armand Hammer and others. They want financial MONOPOLY. Why? Because monopoly guarantees profit.*
>
> *"The New World Order generated by a financial capitalists is WORLD MONOPOLY.*
>
> *"Old John D. Rockefeller once said, 'Competition is a sin.' That credo is supported by the monopolists in New York and the monopolists in Moscow. That's why David Rockefeller and Armand Hammer get along so well with Brezhnev and Gorbachev. They all think power. They all plan for power."*

The recent massive manipulation of the dollar against the yen to try to stimulate the Japanese economy, which is on the rocks, and raise the value of the dollar going into next year's elections shows just how powerful these people are. The banks waited until most currency traders were on vacation and trading was in the summer doldrums, as it usually is in August. Then, in a coordinated effort, they started selling yen and purchasing the dollar. The banks were specifically trying to drive the yen down to the level at which they knew the currency traders would jump on the bandwagon and accelerate the trend at no more cost to the banks. This scheme has worked like a charm so far. This will lower the value of the yen and make Japanese products more competitive in the U.S. Eventually, however, the trillions of dollars we keep pumping overseas will try to come home, as people lose trust in all of the phony money floating around. Then, look out. Head for the financial hills with some gold and silver, if you can.

In other words, the multinational banks are cutting the throats of American businesses and workers, who have to compete against the lower prices of the foreign products, and farmers who have to sell their products at lower prices because of the rising dollar. This is being done to protect the Japanese banks which are in desperate trouble, and to make it possible for the U.S. to lower interest rates, create more instant money out of nothing, and stimulate our economy to try to keep Clinton in office. Unfortunately, this increased money will not go into the consumer economy, but into the speculation bubble, just like it did when the Federal Reserve did the same thing for Bush. Both Clinton and Bush are/were members of the CFR/TLC.

Speaking of the U.N., if it is not enough that the U.S. finances so much of the U.N. already, they now want to get their hands directly in our pocketbooks through a U.N. income tax. If that happens, the U.S. taxpayers will end up paying to forge their own chains, just as our domestic income tax has allowed the U.S. government to nearly explode in size and power.

THE FREEDOM VIEW – THE WAY OUT IS TO GO BACK

"Unless people control government, the government will control them." - Ronald Reagan

We have to build a new major political party, dedicated to honest money and a to federal government that only exercises the powers granted to it by the Constitution. In other words, a Constitution Party. I am convinced that only a party named the Constitution Party and dedicated to restoring constitutional government can ever hope to win the hearts of the American people sufficiently enough to displace one of the two major parties.

The last time a major party arose (Republican), the previous major party (Whigs) had self-destructed and the nation was facing the possibility of a civil war. Because the Democrats split into North and South groups, the Republicans were able to elect Lincoln. The

Democratic Party has not self-destructed, at least not yet. The Republicans are not split in half. It will be very difficult to replace one of these parties, but it must be done. Both parties are so entrenched in the power structure and have so many vested special interests groups supporting them that they will never do what needs to be done.

The Libertarian Party shares many of the concerns about big government and individual freedom expressed in this book. They have been around for a long time and have attracted a few votes, most of them protest votes from people who don't really know what they believe. A lot of people have voted for the Libertarian candidate as a vote for "none of the above". The Libertarians have not gotten their message across for a number of reasons. However, if they are willing to use their expertise and join together with others to support a Constitution Party, then they can be very helpful in helping that party change the country.

The Tax Payers' Party also shares many of the values expressed in this book. They are hampered by a one-issue name which will prevent them from reaching permanent prominence no matter how mad people are about taxes. If they will join together with others in a Constitution Party, then the bandwagon will begin to roll.

Ross Perot can never be elected President, particularly after the fiasco of the 1992 elections. He did show how many disaffected people are willing to work for anybody who can mount a viable campaign. If he, and/or most of his followers, will join together with others in promoting a Constitution Party, we will be well on our way.

There are many other small parties and groups that put forward candidates from time to time. These have no chance for a variety of reasons, but they could help build a Constitution Party.

The reason I believe that only a Constitution Party has any chance is that people can immediately recognize what it stands for and identify with it. Only it would have the philosophical foundation to become one of the major parties. Protest votes may muddle a national election, but they can't win one. Parties built on one man or his views or money can never become a serious, permanent party.

In order to become a permanent major party, it (we?) will have to have candidates for every office, from precinct committeeman to President. Hopefully, these candidates will not have been previously elected to office as Republicans or Democrats, or are totally disgusted with the two parties and be willing to commit to restoring constitutional government. The Constitution Party will have to have workers who are willing to pass out literature, make phone calls, provide rides to polling places, contribute lots of money as individuals, and do all the other things that a major party has to do.

What are you willing to do? How about running for an office? You will have to compensate for the money spent by the two major parties on ads with lots of work including going door to door in local elections.

In order to be taken seriously as a major party, the Constitution Party will have to have a nominating convention just like the other two at about the same time. It will need to have elected delegates to choose the candidates in a democratic manner. It also will have to convince the American people that the party is not just a vehicle for some man to make a run at the Presidency, and the organization will fade out of sight when he doesn't win.

Forget about reforming one of the other two parties. The CFR/TLC crowd will never let it happen. They own most of the leaders in both parties. This has to be a grassroots movement and must be fueled by the money, time, and work of millions of people if it is to be successful. The movement must be organized in an extremely short period of time. While few people will agree with me, I am convinced from the tide of history that if the Constitution Party does not win the presidency or elect sufficient members of Congress to hold the balance of power between the other two parties in 1996, we will not have another chance.

Anybody want to organize a Constitution Party? How about the Tax Payers' Party changing their name and broadening their philosophical base? Any other suggestions out there? Somebody has to do something if we are to wage a real revolution at the ballot box. Make that a lot of somebodies. The only alternatives appear to be serfdom in the New World Order or a revolution of bombs and bullets, which would destroy the country.

Chapter Eleven

The Declaration of Independence

"In Europe, charters of liberty have been granted by power. America has set the example and France has followed it, of charters of power granted by liberty (people-RT). *This revolution in the practice of the world may, with an honest praise, be pronounced the most triumphant epoch in its history and the most consoling presage of its happiness."* - James Madison.

The Declaration of Independence is the charter of our nation. It sets forth the basic ideas and ideals upon which the new nation was to be created. While the Constitution provides the specifics of government, the Declaration outlines the underlying principles. When confusion arises about the intent of a particular item or phrase in the Constitution, the courts should and do sometimes refer to the underlying principles enumerated in the Declaration. At the least, no interpretation or decision should be rendered that is in violation of these principles since the same people wrote/approved both.

"The Declaration of Independence was formed by the representatives of American liberty from thirteen states... Now, my countrymen, if you have been taught doctrines which conflict with the great landmarks of the Declaration of Independence, if you have listened to suggestions which would take from its grandeur, and mutilate the symmetry of its proportions...let me entreat you to come back... Do not destroy that immortal emblem of Humanity, the Declaration of Independence." - Abraham Lincoln

Declaration of Independence
(Adopted in Congress 4 July 1776)

The Unanimous Declaration of the Thirteen United States of America....

These "united" states were separate, sovereign states who were uniting for the purpose of a common cause, namely, the overthrow of tyranny and the gaining of their independence. They retained almost all of the rights of sovereign nations and exercised those rights under the Articles of Confederation.

We hold these truths to be self-evident, that all men are created equal, that they are endowed by their Creator with certain unalienable rights, that among these are life, liberty and the pursuit of happiness.

No earthly power had decreed the truths that Jefferson was about to enumerate. They were self-evident. No reasonable person could effectively argue against them. No one had been given the power to make himself the ruler over others, for no one had any innate right to take away the rights of others. This statement was a direct challenge to the so-called divine right of kings.

Humans are endowed by their Creator with certain inalienable rights. No earthly government had the right to grant or withhold these rights, for they were granted by the One who was above all governments. This is one area where evolution has undercut the foundations of society. If we are nothing but animals, and if the basic principle of evolution is the survival of the fittest, then as many evolutionists themselves have admitted, wars are the natural outcome of man's evolution. Domination of others by one man or by a group of men is the inevitable result. There are no rules. There are no inherent rights. It is every man for himself. To the strongest belong the spoils.

Jefferson enumerates three basic rights, but did not intend to limit the number of basic rights to just these three described below:

life, Everyone has the right to live, to protect himself (that is, self-defense), and this right cannot be arbitrarily taken from him. While the state has a right to impose death as a penalty for violating someone else's right to life, and protect others from similar violations by the same person, it cannot do so without establishing by due process the guilt of the party and making sure that the rights of the accused are protected.

In 1973, the Supreme Court struck down all restrictive state laws against abortion. The court said that an individual had the right to do with her own body whatever she wanted. The individual's body did *not* belong to the *state*. In 1980, the Supreme Court refused to hear a case that would have allowed a terminal cancer patient the right to use Laetrile to treat the disease. The court stated that the individual's body did *not* belong to the *individual*, but to the *state*, and that the state had the right to tell the individual what he or she could do with their own body.

Confused? I am. If this isn't schizophrenia (insanity), what is it? The 1980 ruling directly contradicted the 1973 ruling, using the same words.

Just as silly from a legal standpoint is the situation where a woman may pay someone to brutally kill her baby in the most painful ways possible or do it herself. This is perfectly legal, since the baby isn't defined as a human being. However, if she takes drugs while pregnant, she can be prosecuted for delivering a controlled substance to a *minor*! A minor *what*?

It is no wonder Justice Sandra Day O'Conner has stated that Roe v. Wade was on a collision course with itself. Unfortunately, she and most of the court have carefully avoided dealing with the contradictions within the decisions. The court has also ignored the destruction of human life resulting from the absence of due process for the baby.

In the Roe v. Wade decision, the Supreme Court said that there was no consensus on when human life began, so they arbitrarily set up the ridiculous trimester system as a compromise. It was doomed from the start, being illogical and unworkable. In effect, the court tried to legislate when it had no experience or authority to do so. This function is clearly left up to Congress and the various state legislatures. The Supreme Court has repeatedly tried to legislate remedies to perceived social problems that Congress had enough sense to stay away from, particularly since the Warren Court. This is a usurpation of power and represents a breakdown of the checks and balances in our federal system.

No one could be sure when one trimester ended and another began. So, how could one set of actions be allowed in one trimester and not in another. The lower courts quickly recognized the silliness of this "decree" and they essentially eliminated the restrictions in the later trimesters.

Even very liberal constitutional lawyers, who believe in abortion as a right, have said that Roe v. Wade and its companion decision were very poor constitutional laws. The "right" to abortion was essentially created out of thin air and is found nowhere in the Constitution. The court's argument, if you can call it that, was based on the Fourteenth Amendment quoted below.

> *"Amendment XIV*
>
> *Section 1. All persons born or naturalized in the United States, and subject to the jurisdiction thereof, are citizens of the United States and of the State wherein they reside. No State shall make or enforce any law which shall abridge the privileges or immunities of citizens of the United States; nor shall any State deprive any person of life, liberty, or property, without due process of law; nor deny to any person within its jurisdiction the equal protection of the laws."*

Read that carefully. Even using a microscope, you can't find the so-called right to privacy (which the government violates every day) that was used to justify abortion. The opening phrase of the amendment does not exclude the unborn, but only addresses the question of who is a citizen. Note that the second part of the second sentence says that the state (government) will not deprive any person (not citizen) of the right to life, liberty or property without due process of law. Yet, that is just what the court did to the babies. There certainly is no due process for them. Furthermore, note that the state cannot deny to *any* person within its jurisdiction the equal protection of the laws. Yet the court barred the states from providing that protection to the babies. The very amendment they tried to use to justify abortion clearly says that you cannot take someone's life without due process. Yet, they struck down the laws of all the states that provided protection for those lives and the necessary due process! Does this make any sense to you?

Schizophrenia again. Court Legislation again. Political correctness again. A "flexible" Constitution (according to Justice Marshall) that the court can make mean whatever is convenient or popular at the time, is no Constitution at all. In clear violation of the very principles upon which this nation is based, they took away the babies' right to life, liberty and the pursuit of anything. They also violated the Eighth Amendment against cruel and unusual punishment. See the next chapter on the Constitution.

> *"A good and faithful judge prefers what is right to what is expedient."* - Horace

With our current understanding of genetics, there is almost universal agreement among experts that a unique human life begins at conception. The genetic code is entirely fixed. Nothing can stop a human being from being born, except some sort of intervention or interference in the normal process. A unique human life has begun. There is no point after conception where you can seriously argue that the baby is not human one minute and is human the next. This has led to the rank absurdity that you can open the womb by caesarean section and kill the baby before it can breath. If the baby takes a breath, however, killing it is murder. What nonsense. The baby is already breathing through the umbilical cord. Just as silly, you can kill a baby that is sliding down the birth canal, but once it emerges, killing it is murder. Such are the absurdities generated by these kinds of arbitrary distinctions.

With the current consensus that life begins at conception, the court's excuse for allowing babies to be arbitrarily and brutally killed and deprived of their "unalienable" right to life falls apart, and is clearly revealed to be absurd. The most fundamental function of government is to protect the lives and property of its people. Our government has clearly abandoned this function in numerous ways.

Once before in history, the Supreme Court made a similar decision. In the Dred Scott case, the court decreed that slaves were not citizens in a ridiculous display of going along with the political correctness of the day. You would have to be a complete idiot to argue such a thing today. The attempt to determine how "human" a baby was in the trimester scheme was absurd. When does the baby *become* human? What utter nonsense. Eventually, the Dred Scott decision was overturned and recognized as absurd. The same thing must eventually happen to Roe v. Wade. If anyone, regardless of age, can have their rights to life, liberty, and the pursuit of happiness arbitrarily terminated without due process, then the same can happen to any of us. The elderly and those whose "quality" of life is deemed to be below par are next on the list. *If* you don't think that due process is extremely important, talk to the relatives of those whom the BATF and the FBI have eliminated without any due process.

liberty, The ideas of the Writ of Habeas Corpus, the right to a jury trial, and the right to do as you please as long as you do not thereby deny someone else their same rights, are a substantial part of the foundation of our country. The Revolutionary War was fought for these liberties. They are what millions of Americans have given their lives to defend. It is what Bill Clinton and the Republican Congress are taking from us in some of the recent and proposed legislation. The provisions of these laws, like that which allows Clinton to designate anyone he wants to a terrorist, suspending all the "terrorist's" Constitutional rights without any due process at all, and to then throw them in jail or the concentration camps without any court review or appeal, are not only unconstitutional, but they violate the basic principles of the country. Anyone who votes for such laws should be defeated at the next opportunity as having violated their oath of office, the public trust, and trampled on the Constitution, regardless of what else they may stand for. If these officials will throw away your most basic rights, they cannot be trusted in any position of power.

> *"Posterity - You will never know how much it has cost my generation to preserve your freedom. I hope that you will make good use of it."* - John Quincy Adams

pursuit of happiness. Note that Jefferson did not say just happiness, (or financial security, or food or whatever), but the *pursuit* of happiness which requires complete freedom of action that doesn't trample on someone else's rights.

That to secure these rights, governments are instituted among men, deriving their just powers from the consent of the governed.

The purpose of government is not to supply the people with anything, but rather to protect them from anyone, internal or external, who would try to deny them what are already their inalienable rights. Governments create nothing. They only "have" what they take away from the people.

America's unique approach to government is that the people have all of the power and the rights and the government can only use what the people choose to relinquish to it. While that basic concept is still given lip service by our politicians, in practice it has pretty well gone down the drain as the courts, Congress and particularly the bureaucrats blithely ignore the Constitution and do as they please.

That whenever any form of government becomes destructive to these ends (as ours has-RT), **it is the right of the people to alter or to abolish it, and to institute new government, laying its foundation on such principles and organizing its powers in such form, as to them shall seem most likely to effect their safety and happiness.**

That our current government has and is trampling the Constitution and ignoring the rights of the people is beyond question. The question is, What should we do about it? We have the *right* to alter or even abolish that government. Fortunately, it is not the form of government that is at fault, but the people in it and their belief in a socialist, nanny government which is so antithetical to the principles on which this country was founded. We can change that if we have the will. All we have to do is vote for people who believe that the Constitution means what it says and says what it means. This new breed would truly represent the people, not PACs, multinational corporations, or groups like the CFR/TLC that so many of the current politicians belong to. Unfortunately, we will have to form a new, major political party dedicated to that end, as difficult as that may be. Most of the politicians in the current two major political parties have proven that they will not honor their oath of office to uphold and defend the Constitution of the United States.

Prudence, indeed, will dictate that governments long established should not be changed for light and transient causes; and accordingly all experience hath shown that mankind are more disposed to suffer, while evils are sufferable, than to right themselves by abolishing the forms to which they are accustomed.

Jefferson was quite right, as usual. You don't change the form of government without a very significant reason. We have no need of doing that. This is what David Rockefeller, the CFR, and the various foundations are trying to do with their New States Constitution in order to bring about a New World Order. We must fight tooth and nail against any tampering with the Constitution. It has served us well and has provided the framework for building the most free and most prosperous nation in history. All we need to do is change the people in power who are ignoring the Constitution. Jefferson also was right in noting that we tend to drift along with what we have, even if it involves suffering the slow destruction of our rights, rather than get up off the couch and trying to do something about it before it is too late.

But when a long train of abuses and usurpations, pursuing invariably the same object evinces a design to reduce them under absolute despotism, it is their right, it is their duty, to throw off such government, and to provide new guards for their future security.

Some of those abuses have been detailed in this book. Many others are spelled out in the books and newsletters listed in the Appendix. The object that is being pursued is a socialistic government that controls every aspect of everyone's life for the benefit of those in power. This directly contridicts the principles on which this country was founded. The federal government has usurped the powers reserved for the states and the people by the Tenth Amendment. The courts and the executive branch have usurped or been improperly given the powers granted to

Congress. The checks and balances built into the system for our protection are largely gone or ignored.

– Such has been the patient sufferance of these colonies; and such is now the necessity which constrains them to alter their former systems of government. The history of the present King of Great Britain is a history of repeated injuries and usurpations, all having in direct object the establishment of an absolute tyranny over these states. To prove this, let facts be submitted to a candid world.

– Such has been the patient sufferance of these states and people, largely bought off by your tax money, and such is now the necessity which constrains them to alter the unconstitutional imbalance of power between the states and the federal government. The history of the federal government since 1932 is a history of repeated injuries, particularly the rampaging law enforcement agencies, and usurpations, all having as their direct object the establishment of an absolute tyranny over these states and people. To prove this, let the facts in this book and those references listed in the Appendix be submitted to reasonable men.

He (the king) **has erected a multitude of new offices, and sent hither swarms of officers to harass our people, and eat out their substance.**

Some things never change. The Agriculture department had one bureaucrat for every 1800 farms in 1900. Now they have one for every 16. Farmers can no longer can plant what they want where they want to. The whole bureaucratic structure continues to grow and feed on its host like some hideous malignant cancer. Eighty percent of the welfare appropriations go not to the needy but to the bureaucrats. Only by slashing the federal government and returning it to its constitutional powers can we remove most of the cancer.

> *"If we were directed from Washington when to sow and when to reap, we should soon want for bread."* - Thomas Jefferson

He has kept among us, in times of peace, standing armies without the consent of our legislature.

This was the reason for the tremendous debate about standing armies that raged during the drafting of the Constitution. The founders feared the use of government troops against the people. It is the basis of our Posse Comitatus laws forbidding the use of the military against citizens which is now being violated repeatedly as in Waco.

He has affected to render the military independent of and superior to civil power.

In our present situation, the government is attempting to establish federal law enforcement agencies as paramilitary assault units, and train them with military units in violation of Posse Comitatus and establish the whole concept of armed federal units to be used directly against the people as in Waco, Ruby Ridge, and far too many other incidents. Clinton and the CFR gang want to establish a national police force for this purpose. This force will not be superior to the civil powers, but used by them to enforce their unconstitutional decrees, such as all of those executive orders to be discussed later.

He has combined with others to subject us to a jurisdiction foreign to our constitution, and unacknowledged by our laws; giving his assent to their acts of pretended legislation:

If this statement isn't a exact description of the provisions of NAFTA and GATT, I don't know how you could get more accurate.

For protecting them, by mock trial, from punishment for any murders which they should commit on the inhabitants of these states:

We have substituted internal investigations by the accused agencies for mock trials, but the results are the same. People with badges murder innocent citizens or threaten them with harm and even brag about their ability to do so without expecting any consequences.

For imposing taxes on us without our consent:

If most of the income tax increases had been submitted to a popular vote, they would have been resoundingly defeated, as are most votes on proposed property tax increases. Yet, we keep sending the politicians back to Congress who vote for these increases. Many of these Congressmen are from districts that largely feed off of the taxpayers, or they are supported by special interests groups that want special treatment and provide the money for them to stay in office. Most of these taxes support unconstitutional expenditures in the first place. The Sixteenth Amendment needs to be overturned or repealed.

For depriving us of the right to trial by jury:

That is where recent laws are heading, particularly the Omnibus Anti-terrorism Act which specifically authorizes such actions.

For taking away our charters, abolishing our most valuable laws, and altering fundamentally the forms of our governments:

– For overriding our state and local laws in violation of the Tenth Amendment, usurping the powers reserved for the states and the people, and declaring unconstitutional state and local laws protecting the health and safety of our people.

All three branches of the federal government have repeatedly violated both the Constitution and the basic principles that this nation was founded upon. Is it time for a new Declaration of Independence? As you can see from the above examples, ample grounds exist. Fortunately, we don't need to declare independence, just reestablish our constitutional government by electing people dedicated to the Constitution and these basic principles.

Chapter Twelve

The Constitution of the United States

"We have given them a republic, if they can keep it." - Ben Franklin

"The history of liberty is a history of the limitations of government power, not the increase of it." - Woodrow Wilson

"Our Constitution is the document that protects the people from the government." - Ronald Reagan

THE CONSTITUTION OF THE UNITED STATES

We the People of the United States, in Order to form a more perfect Union, establish Justice, insure domestic Tranquility, provide for the common defence, promote the general Welfare, and secure the Blessings of Liberty to ourselves and our Posterity, do ordain and establish this Constitution for the United States of America.

We the people... No state, nation, or other organized group had the right to form a government to rule over the people. Only the people had the power, in keeping with the principles set down in the Declaration of Independence, to delegate a limited amount of that power to any government. That power was to be specific and strictly limited.

...form a more perfect union, The Confederacy just didn't work. There was not enough cooperation among the states.

...establish justice, Criminals could flee from one state to another and usually be safe. Laws varied widely.

...insure domestic tranquility, This was to prevent trade wars, border disputes and potential armed conflict between the states.

...provide for the common defense, A real problem was trying to get other states to help when one or two were attacked.

...promote the general welfare, This primarily referred to duty-free trade and increased commerce. It had nothing to do with any specific individual's welfare.

... secure the blessings of liberty... The Constitution has done a superb job of doing this, which is why the CFR/TLC crowd wants to replace it or make major changes. As Ben Franklin said, "We have given them a republic, if they can keep it." Can we? It is under very serious attack.

Section 1. All legislative Powers herein granted shall be vested in a Congress of the United States, which shall consist of a Senate and House of Representatives.

You can't get much clearer than this. Yet, both the courts and the executive branch are increasingly exercising legislative powers. The executive branch does so through Presidential executive orders and all of those bureaucratic regulations.

Section 8. The Congress shall have Power To lay and collect Taxes, Duties, Imposts and Excises, to pay the Debts and provide for the common Defence and general Welfare of the United States; but all Duties, Imposts and Excises shall be uniform throughout the United States;

Note that this power to lay taxes was severely restricted by a later section. The money was to be used for the common defense and *general* welfare, not for the welfare or retirement scheme of various individuals, nor for income redistribution. The words in the Constitution were very carefully chosen with a great deal of discussion over them. "General" meant the overall common welfare. No authority was granted to provide support for one individual at the expense of another. That scheme would have been inconceivable to the authors.

To borrow Money on the credit of the United States;

Since the government has shown itself to be totally irresponsible, this needs to be repealed or severely restricted.

> *"I wish it were possible to obtain a single amendment to our Constitution. I would be willing to depend on that alone for the reduction of the administration of our government to the genuine principles of its constitution; I mean an additional article, taking from the federal government the power of borrowing."* - Thomas Jefferson

Again, Jefferson is correct. It is long past time for his amendment to be passed.

To regulate Commerce with foreign Nations, and among the several States, and with the Indian Tribes;

Note that the Indian tribes had the standing of foreign nations. Trade was to be regulated among the several states, but with foreign nations and Indian Tribes.

To coin Money, regulate the Value thereof, and of foreign Coin, and fix the Standard of Weights and Measures;

Note that this section has *not* been repealed. Granting the power to privately owned banks to manipulate our money supply and value was anathema to the founding fathers. Most of them fought against Hamilton's attempts to form a central bank. Granting such a bank, or collection of banks such power would be unthinkable to most of the founding fathers. Yet, that is just what Congress did in 1913, and we have been paying through the nose to them ever since. The Federal Reserve Act of 1913 must be repealed and this country must return to an honest money system where money has to be earned, not created out of thin air as the banks do now.

To establish Post Offices and post Roads;

This clause grants the Congress the right to establish post offices, in order to promote the general welfare, as no one else had the capital to do it. Congress was not given the authority to limit this service exclusively to government post offices.

To promote the Progress of Science and useful Arts, by securing for limited Times to Authors and Inventors the exclusive Right to their respective Writings and Discoveries;

Note that this refers to the promotion of the progress of science through copyright/patent protection. No authority is granted to Congress to spend money on supporting the *production* of such items, certainly not obscene paintings.

To constitute Tribunals inferior to the Supreme Court;

Congress can set up these courts and define what they can and cannot do.

To raise and support Armies, but no Appropriation of Money to that Use shall be for a longer Term than two Years;

The rather strange restriction on the appropriation of money was to address the fears of many that a standing Army would be used against the citizens of the country, which is now being done in violation of the posse comitatus laws.

To provide and maintain a Navy;

Note that there was no such restriction on appropriations for the Navy, as it could attack few citizens. That was before the Marines and the Navy Seals.

To make Rules for the Government and Regulation of the land and naval Forces;

The Constitution of the United States - 241

To provide for calling forth the Militia to execute the Laws of the Union, suppress Insurrections and repel Invasions;

To provide for organizing, arming, and disciplining, the Militia, and for governing such Part of them as may be employed in the Service of the United States, reserving to the States respectively, the Appointment of the Officers, and the Authority of training the Militia according to the discipline prescribed by Congress;

This clause assumes a volunteer citizen militia and provides authority for working with it, but *restricts* the power of the Congress to govern or regulate the militia except to that part which shall be *employed* by the government. To further protect the people from the misuse of the military, the appointment of the officers and any training of the militia in a state was under the control of the individual states.

To exercise exclusive Legislation in all Cases whatsoever, over such District (not exceeding ten Miles square) as may, by Cession of particular States, and the Acceptance of Congress, become the Seat of the Government of the United States, and to exercise like Authority over all Places purchased by the Consent of the Legislature of the State in which the Same shall be, for the Erection of Forts, Magazines, Arsenals, dock-Yards, and other needful Buildings; – And

The federal government could *purchase* land in the states, but only with the consent of that state's legislature. The land originally belonged to the state, or to the persons to whom the land was sold by the state.

The Privilege of the Writ of Habeas Corpus shall not be suspended, unless when in Cases of Rebellion or Invasion the public Safety may require it.

This right of proper due process when detained, along with the rights to a trial by jury, and the right of self defense are the foundation of all other personal rights. If these aren't there, none of the others matter very much. When these two rights were wrested away from the King earlier in English history, the monarchy's power over the people was largely broken. The Habeas Corpus right is being eliminated, both by executive orders as well as recent legislation.

No Bill of Attainder or ex post facto Law shall be passed.

Until Congress passed Bill Clinton's retroactive income tax increase, which is blatantly unconstitutional.

> *"June 13, 1994 will go down in the history of the demise of our country as the day when property lost all protection against retroactive taxation. Those who look to the Supreme Court to protect Constitutional rights look in vain. The only way we can protect ourselves from dispossession is to carefully monitor the legislative action of those we elect."* - Paul Craig Roberts, former Under Secretary of Treasury

How about electing honest people to the legislatures who will not violate their oath of office?

No Capitation, or other direct, Tax shall be laid, unless in Proportion to the Census or Enumeration herein before directed to be taken.

There were very good reasons for this strong restriction against levying direct taxes on individuals. The government was restricted from impoverishing the people and the amount of credit the government could obtain was likewise limited. We threw away this protection with the Sixteenth Amendment, which was never legally passed, and we need to overturn it or repeal it.

No Money shall be drawn from the Treasury, but in Consequence of Appropriations made by Law; and a regular Statement and Account of Receipts and Expenditures of all public Money shall be published from time to time.

This restriction is violated every day as there are all kinds of funds dispersed by various schemes that are not included in direct appropriations bills, largely due to unconstitutional laws setting up activities that the federal government has no authority to set up in the first place.

Section 10. No State shall enter into any Treaty, Alliance, or Confederation; grant Letters of Marque and Reprisal; coin Money; emit Bills of Credit; make any Thing but gold and silver Coin a Tender in Payment of Debts; pass any Bill of Attainder, ex post facto Law, or Law impairing the Obligation of Contracts, or grant any Title of Nobility.

The states reserved for themselves the right to make payment of debts in gold and silver coin legal. They may end up doing that to promote commerce when fiat paper money becomes worthless. Having reserved that right, which the federal government cannot restrict, the confiscation of gold coins by Roosevelt was clearly unconstitutional, since any state had the unrestricted right to declare them legal tender.

Section 4. The President, Vice President and all civil Officers of the United States, shall be removed from Office on Impeachment for, and Conviction of, Treason, Bribery, or other high Crimes and Misdemeanors

Article III

Section 1. The judicial Power of the United States, shall be vested in one supreme Court, and in such inferior Courts as the Congress may from time to time ordain and establish. The Judges, both of the supreme and inferior Courts, shall hold their Offices during good Behavior, and shall, at stated Times, receive for their Services, a Compensation, which shall not be diminished during their Continuance in Office.

Supreme Court justices do not have a guaranteed lifetime appointment. They can be impeached for bad behavior.

In all Cases affecting Ambassadors, other public Ministers and Consuls, and those in which a State shall be Party, the supreme Court shall have original Jurisdiction. In all the other Cases before mentioned, the supreme Court shall have appellate Jurisdiction, both as to Law and Fact, with such Exceptions, and under such Regulations as the Congress shall make.

Congress has very wide latitude in dealing with the courts. The judicial branch was deliberately designed to be the weakest because judges were not elected and were, historically, the source of much abuse. Congress can regulate, restrict and even eliminate all courts except for the Supreme Court. It can also severely limit the authority and scope of the Supreme Court itself. It may be time to do that in order to restore a semblance of the intended balance of power between the branches of government.

The Trial of all Crimes, except in Cases of Impeachment, shall be by Jury; and such Trial shall be held in the State where the said Crimes shall have been committed; but when not committed within any State, the Trial shall be at such Place or Places as the Congress may by Law have directed.

This is one of our most fundamental rights. There is a move afoot to limit which crimes *deserve* a trial by jury because the courts are overloaded. That would take a constitutional amendment unless the Supreme Court wimps out, as it often does, for the sake of convenience or political correctness. It would be much better to get rid of millions of unconstitutional laws and regulations and reform the justice system. That would help to clear out the criminal courts quickly.

Article IV

Section 1. Full Faith and Credit shall be given in each State to the public Acts, Records, and judicial Proceedings of every other State. And the Congress may by general Laws prescribe the Manner in which such Acts, Records, and Proceedings shall be proved, and the Effect thereof.

Congress may act to facilitate how the states recognize each other's laws, but it is not given the authority to modify or overturn those laws, which it has repeatedly done. Remember, these were sovereign states ceding certain very specific, limited powers to the central government and retaining *all* others.

Section 4. The United States shall guarantee to every State in this Union a Republican Form of Government, and shall protect each of them against Invasion; and on Application of the Legislature, or of the Executive (when the Legislature cannot be convened) against domestic Violence.

Only the state legislature may request that federal troops be used for cases of domestic (internal) violence in their state if they are in session – not the governor, mayor, or anyone else. The governor can make the request only if the legislature is not in session. In addition, federal troops could only be used in cases of invasion or domestic violence. This provision was intended to protect the states and their citizens from having federal troops used against them. This has been violated in the past and the violations are growing rapidly.

Article V

The Congress, whenever two thirds of both Houses shall deem it necessary, shall propose Amendments to this Constitution, or, on the Application of the Legislatures of two thirds of the several States, shall call a Convention for proposing Amendments, which, in either Case, shall be valid to all Intents and Purposes, as Part of this Constitution, when ratified by the Legislatures of three fourths of the several States, or by Conventions in three fourths thereof, as the one or the other Mode of Ratification may be proposed by the Congress; Provided that no Amendment which may be made prior to the Year One thousand eight hundred and eight shall in any Manner affect the first and fourth Clauses in the Ninth Section of the first Article; and that no State, without its Consent, shall be deprived of its equal Suffrage in the Senate.

The convening of a new constitutional convention, as Rockefeller and his cohorts are trying to do, is *not* authorized by the Constitution. A convention may be called *only* for the purpose of *suggesting* amendments, and those must be ratified by three-fourths of the states. It will be interesting to see if these people do manage to call a convention in order to ram their Newstates Constitution down our throats, ignoring the restrictions in this provision. They plan on pretending that a convention to suggest amendments can declare itself a constitutional convention. Then, it could declare itself a convention of the states who have sent representatives (hand picked by the CFR) and ratify its own suggestions. However, the passage above says that the amendments must be ratified by conventions *in* the respective states, not by a convention of delegates *from* the states. It further states that Congress has the power to decide which method is to be used to ratify anything. None of these restrictions will stop them from trying it their way anyway.

This Constitution, and the Laws of the United States which shall be made in Pursuance thereof; and all Treaties made, or which shall be made, under the Authority of the United States, shall be the supreme Law of the Land; and the Judges in every State shall be bound thereby, any Thing in the Constitution or Laws of any State to the Contrary notwithstanding.

This clause was inserted to make sure that the Constitution, federal laws, and treaties were enforced by the states. Note that the laws must be made in accordance with the Constitution. That is, federal laws were valid only if they conformed to the specific powers and authority granted to the federal government. Some have interpreted this section to mean that treaties with foreign governments override state and even federal laws and must be enforced in the states. This is true *only* if the treaties also conform to the Constitution's restrictions and to the authority granted the federal government, that is, Pursuance thereof. Some have even interpreted this section to mean that treaties could override the Constitution. This is not true. The authors of the Constitution made *no* provision whereby *anyone* could override the Constitution. Treaties must be made under and in accordance with the

Constitution. *Nothing* can override the Constitution, which is why all of those executive orders that suspend part or all of the Constitution are, by definition, unconstitutional. Yet, they remain on the books and must be rescinded before someone tries to activate them. If they are used, this will lead to open civil war in this country, and rightfully so under the provisions of the Declaration of Independence.

The Senators and Representatives before mentioned, and the Members of the several State Legislatures, and all executive and judicial Officers, both of the United States and of the several States, shall be bound by Oath or Affirmation, to support this Constitution; but no religious Test shall ever be required as a Qualification to any Office or public Trust under the United States.

What a sorry joke this requirement has become.

AMENDMENTS TO THE CONSTITUTION OF THE UNITED STATES

The first ten amendments are known as the Bill of Rights and were inserted at the insistence of many of the states as a prerequisite for ratification. They were primarily designed to protect the "inalienable" rights of individuals against infringement by the government.

> *"..a bill of rights is what the people are entitled to against every government on earth."* - Thomas Jefferson.

> *"The right to be left alone is the underlying principle of the Constitution's Bill of Rights."* - Dr. Erwin N. Griswold, Dean of the Harvard School of Law.

I wonder how Randy Weaver and the survivors of Waco feel about their bill of rights?

Amendment I

Congress shall make no law respecting an establishment of religion, or prohibiting the free exercise thereof; or abridging

the freedom of speech, or of the press; or the right of the people peaceably to assemble, and to petition the Government for a redress of grievances.

Boy, has this amendment been abused and misinterpreted. Whole books have been written about the "establishment" clause alone. In the simplest terms, there is not one word in the entire Constitution about a "wall of separation" between church and state. That phrase came from an offhand comment by Jefferson. The Constitution says that *Congress* shall make no law regarding the establishment of religion. First of all, the restriction did not apply to the states. Extending the prohibition to the states was thoroughly discussed and rejected by the delegates. The Supreme Court has extended it to the states anyway by judicial fiat, but such a restriction is not in the Constitution. It should be, but it isn't. If we want it there, it should be by amendment, not the desires of five or even nine individuals.

Second, virtually all of the founding fathers believed that religious convictions were necessary for the governing of any people in a republic. The federal government was restrained from establishing any religion as the government supported national religion which was the case in most other countries. The Constitution actually says nothing about favoring any particular religion, particularly in the states, which was permissible and was frequently done. I think that today, neither the federal government nor the states should favor any one religion, but this could be accomplished by Congress even without an amendment. Certainly, the founders never intended the government to be hostile to religion or to bar local communities, school boards, or anybody else from participating in religious celebrations or teaching general religious values. Strangely enough, the *courts* have established one particular religion, humanism, as the state religion, and it is the only one which can be taught in the public schools. They have even specifically designated humanism as a recognized religion. But the courts still approve of teaching the basic tenets of humanism under the guise of science, and forbid the teaching of contending concepts despite the fact that these other concepts are based on far better scientific and historical evidence.

Both the courts and Congress have repeatedly violated the second part of the first sentence cited above by prohibiting or restricting the free exercise of religion. The whole sentence was there to prevent the government from interfering with the free exercise of religion either by establishing a state religion or passing laws (or court decrees) that affected free exercise in any way or in any place. Yet, they now do so frequently.

The EEOC proposed regulations that people could not have Bibles on their desks or wear jewelry at work that had any religious indication. Fortunately, these proposed regulations have been put on the shelf for now. Yellow Pages publishers began removing Christian symbols from phone directories because of pressure from self-declared lesbian Roberta Achtenberg, who is director of HUD's Fair Housing Division. People who paid for these ads have every right to the free expression of their religious persuasion or identities, but forget the Constitution if someone isn't politically correct. Many of the people in our government will do anything to suppress religion, particularly Christianity, which they hate.

... or abridging the freedom of speech, This prohibition is ignored in the case of politically incorrect speech. It is fascinating to see liberal academia trying to enforce politically correct speech on students. Whatever happen to real liberalism? Both Bill Clinton and the previous Democratic Congress were trying to find ways to muzzle the free speech of their opponents, particularly religious or talk radio shows. The neo-liberals believe in free speech only for themselves. Now, the Omnibus Anti-Terrorist Act gives Clinton the power to jail or silence anyone he chooses by designating someone who criticizes him a terrorist and stripping him of all his constitutional rights.

... or of the press, According to an article in the Orange County Register entitled: *"The Press Admits They Have Been Ordered to Cut Off Coverage"*, the mainstream media have been told to censor investigations of Clinton criminal misconduct, promiscuity, and scandals. Fortunately, the British press has no such pressure and has printed massive exposes of Clinton's conduct. Although the stories are filled with specific facts, verified by many witnesses, none of it has been reprinted or broadcast in the mainstream U.S. media. Isn't it odd that people in other countries know more about the Clinton scandals that we do?

... or the right of the people to peacefully assemble, You can assemble, picket, and demonstrate unless you are assembling, picketing or demonstrating in front of an abortion clinic. Then, you can be brutally beaten by the police and/or thrown in jail for very long terms with heavy fines. The protestors against killing babies are receiving the same kind of treatment that the civil rights protestors received in the beginning of the civil rights movement and it is just as illegal.

... and to petition the Government for redress of grievances, Attention, Congress! Consider this book a petition for redress of grievances that demands: compensation for of all the people trampled on by the bureaucracies gone berserk; the cessation of stealing money by an income tax that was never properly ratified; the return of all the money confiscated by a Ponzi scheme that was not put in trust, as promised, but squandered by a lying government; and the repeal of all the unconstitutional laws and regulations that burden and harass a people whose freedom from such harassment is guaranteed by this Constitution. Millions of people are asking for the same thing. Will you respond? If not, you should be replaced.

Amendment II

A well regulated Militia, being necessary to the security of a free State, the right of the people to keep and bear Arms, shall not be infringed.

This one will take a while. I have gone into great detail on this right because the one-world bunch, the Communists, the Socialists, and the neo-liberals have targeted it as being the main impediment to their plans for us peons. This is a primary battleground.

I am not a particular fan of guns. Despite having a real love for the outdoors and writing about the outdoors as a hobby, I do not go hunting as I have no desire to shoot anything. I even gave the shotgun

I inherited from my father to my son. What follows is not a desire to defend a personal preference, but an analysis from history and the Constitution. I would prefer that guns had never been invented. A lot more people would be alive today. However, they were invented and we have to deal with the fact that the person or group that has the most/best/biggest guns usually rules over those who don't.

A gentlemen named Neil Schulman asked one of the best known linguistic experts on English usage in the world (Prof. Roy Copperud) to analyze the wording of the Second Amendment. It was a kind of double blind analysis. He had no idea of the professor's political or constitutional views, and likewise, Prof. Copperud had no idea of Schulman's views.

Summarizing the questions by Schulman and the answers and analysis by Copperud (fairly I hope), Prof. Copperud indicated that: 1. The language assumes the right of the citizen to bear arms (one of those inalienable rights-RT). The Constitution does not grant that right, as it was pre-existing, but instead recognizes it and forbids the government from interfering with it 2. The part about a militia places no restrictions or requirements on the individual citizen's right to bear arms, whether they are part of any militia or not.

Less there be any doubts about the intent of the authors of the Constitution, let me provide a few quotes.

> George Washington: *"Firearms stand next in importance to the Constitution itself. They are the American people's liberty teeth and keystone under independence. To secure peace, securely and happiness, the rifle and the PISTOL are equally indispensable. The very atmosphere of firearms everywhere restrains evil interference - they deserve a place of honor with all that is good."*

Clinton and company are trying to blame guns for all the bad things, most of which have been caused by government programs or actions in the first place.

> Samuel Adams: *"The Constitution shall never be construed .. to prevent the people of the United States who are peaceable citizens from keeping their own arms."*

Unfortunately, Mr. Adams, that is exactly how the laws are being written.

> Patrick Henry: *"The great object is that EVERY man be armed* (as in Switzerland-RT). *EVERYONE who is able may have a gun...Are we at last brought to such a humiliating and debasing degradation that we cannot be trusted with arms for our own self defense?"*

> Noah Webster: *"Before a standing army or a tyrannical government can rule, the people must be disarmed,- as they are in almost every kingdom in Europe. The SUPREME POWER IN AMERICA cannot enforce unjust laws by the sword; because the whole body of the people are armed, and constitute a force superior to any band of regular troops that can be, on any pretense, raised in the United States."*

> Thomas Jefferson: *"No free men shall ever be debarred the use of arms."* *"The strongest reason for the people to retain the right to keep and bear arms is, as a last resort, to protect themselves against Tyranny in government."*

This is the main reason for the Second Amendment, not hunting.

> James Madison: *"Besides the advantage of being armed, which Americans possess over the people of almost every other nation, the existence of SUBORDINATE governments...forms a barrier against the enterprises of ambition,...the several kingdoms of Europe are afraid to trust the people with arms.*

> Alexander Hamilton: *"Little more can reasonably be aimed at, with respect to the people at large, than to have them properly armed and equipped; ...."*

> Richard Henry Lee: *"To preserve liberty, it is essential that the whole body of the people always possess arms and be taught alike, especially when young, how to use them."*

> George Mason: *"Divine providence has given to every individual the means of self defense .. To disarm the people ... is the best and most effectual way to enslave them."*

...as Clinton and many others are trying to do. Contrast the above statements, particularly the last one, with statements made by our current leaders.

> Janet Reno: *"Waiting periods are only a step. Registration is only a step. The prohibition of private firearms is the goal."*

So much for her oath of office to uphold the Constitution. Her goal is a complete violation of the Constitution and the foundation of our freedom. This is from the person who authorized the all-out assault at Waco rather than continue negotiations even after Koresh had promised to surrender as soon as he finished writing some material he was working on. Obviously, someone didn't want that material finished.

> Bill Clinton: *"We believe very passionately in the Brady Bill. It will be the beginning of what MUST be a long and relentless assault on the problem of crime and violence in the country."*

This means, as others have expressed, that the Brady Bill is the camel's nose in the tent on the way to complete gun confiscation. Dear Mr. Clinton: How about an assault on the government's programs that *encourage* poverty, crime and violence?

Bill Clinton has asked Janet Reno and the Justice Department to study and develop a scheme for national registration and licensing of all handgun owners. Larry Pratt, executive director of Gun Owners of America, pointed out in a *USA Today* editorial that after Wisconsin imposed a 48-hour waiting period, the FBI said that the state's murder rate rose by sixty percent. He wrote, "But there's something even more dangerous - Clinton's gun control program - that is, the registration of gun owners. Clinton is calling for a system that requires gun dealers to notify the government about a potential buyer for a background check. Whenever gun buyers are singled out at a gun store for identification by the government, a list of gun owners can easily be created. This is gun-owner registration. And in places like New York City, police have used gun owner registration lists to confiscate firearms from law-abiding citizens."

What does the word "infringed" mean? This word was chosen very carefully. The people who used it in the Constitution meant that the government was to do *nothing* that restricts the right of people to keep weapons to defend themselves. Read their own writings if you disagree. A short reasonable waiting period, as in the Brady Bill, probably would not be construed as any real infringement by the courts today, although it would have been in their day. However, the Brady Bill is clearly unconstitutional under the Tenth Amendment and has been ruled so by several courts thanks to some courageous sheriffs who don't agree with this destruction of our constitutional rights. Most other gun control laws now on the books, as well as the ones in the pipeline, are even more blatantly unconstitutional, regardless of what our highly politicized courts might say.

The thought/speech/gun control crowd wants to take away everybody's gun to "control crime". At least that is what they claim. That's not the reason. They know better. Removing guns from law abiding citizens does not decrease crime, it increases it. They know the statistics and studies better than the vast majority of people do, and deliberately ignore or distort them.

In parts of Europe where few people have guns, criminals are much more likely to break into a house they think is occupied than in the U.S. The reason is that they are much more likely to get shot in this country, and therefore look for unoccupied houses. Legal restraints on the purchase or possession of guns have virtually no effect on reducing crime. Actually, they help to increase crime.

Only two percent of the 65 million privately owned handguns are used to commit crimes. Only a small percentage of that two percent were purchased legally. The attempt to disarm us has nothing to do with crime and violence and everything to do with government control over our lives.

Attempts to reduce the number of guns available would only reduce the number available for law abiding citizens to defend themselves. Criminals can easily get guns from outside the country if that became necessary. Gun smuggling would become nearly as lucrative as drug smuggling, but our "protectors from ourselves" never seem to learn.

According to one police official, "The police cannot prevent most crimes. They cannot be everywhere at once, not can they anticipate where a criminal will strike next. They can only pick up the pieces, or bodies after the fact....Armed citizens have to protect themselves. We cannot. It is the law of the jungle."

According to Gary Kleck, a criminologist at Florida State University, guns are displayed or fired more than a million times a year by citizens to defend themselves. This outnumbers the total number of arrests (forget convictions) for violent crimes. According to the Justice Department, people who brandish or use a gun for self defense are far less likely to be injured or lose property in an assault than those who don't. This data is from the same Justice Department that wants to take away our guns and leave us defenseless. The claim that people are more likely to hurt themselves or some innocent person is a deliberate lie. Far more lives, not to mention property, are saved by the use of a gun for self defense than are lost by accidents. Only about two percent of gun-related accidents involve the shooting of someone mistaken for an intruder. Most gun accidents involve rifles, shotguns, or other firearms, not handguns. Our real problem is that if you use a gun to defend yourself or your family, you are likely to be punished by the police and/or the courts far more than the criminal you were protecting yourself from. If people do not have the right, which includes the capability, to defend themselves from whomever, then they have no personal or property rights at all. Everything they have, including their lives, can be taken away by anyone who chooses to do so. It is not some statistical aberration that violent crime always goes down when states allow qualified citizens to carry concealed weapons.

Here are just a few of those statistics. Each year, potential crime victims kill between 2,000 to 3,000 criminals. Yet, these private citizens only mistakenly kill about 30 people a year. This is a 100-to-1 ratio. Law enforcement agents, who are far fewer in number, mistakenly kill over 300 people a year. Firearms that are illegally obtained are used to commit about 800,000 crimes a year. On the other hand, legal firearms are used to prevent over 2 million crimes a year. Who wins if the legal guns are taken away? Criminals are three times more likely to be killed by armed victims who resist them than they are by the police. Most felons agree that the main reason burglars avoid houses when people are home is that they fear being shot during the crime.

In Orlando, Florida, a safety course taught Orlando women how to use guns. As a result, the rape rate there dropped eighty-eight percent in 1987. The rape rate stayed constant in the rest of Florida. The criminals went elsewhere, where crime was much safer. From a Justice Department study, it was found that of more than 32,000 recorded attempted rapes thirty-two percent were actually committed. When a woman was armed with a gun or knife, only *three percent* of the attempted rapes were successful. Which group would you rather fall into?

In states where concealed-carry laws have been passed, recognizing the constitutional right of self defense that has been trampled on by federal, state and local governments, violent crime has dropped significantly. When Georgia made it easier for a people to carry a concealed gun, the homicide rate dropped by twenty-one percent, while the national rate rose six percent during the same period.

When Kennesaw, Georgia (a suburb of Atlanta) passed a law requiring heads of households to keep at least one weapon in the house, the residential burglary rate dropped *eighty-nine* percent. Where would you rather live, there or in Washington, D.C., Chicago, or New York City, which have the toughest gun "control" laws in the country? In Washington, DC, with all of that "control", the homicide rate in 1992 was 75.2 per 100,000. In the surrounding urban areas without the harsh controls, the homicide rate was less than 10 per 100,000. Where would you rather live: a place where you have the right to defend yourself, or one where you have to rely on 911 to get someone there in time?

Over eighty percent of our population, which probably means you, will be victims of violent crime at some point in their lives. In any given year, serious crime hits twenty-five percent of all households. Do you want to lose the right to defend yourself like the people in New York City have? In California, gun owners came under immediate police surveillance when some of them were required to register their semi-automatic weapons a few years ago. Most owners refused to do so. As Don McAlvany has pointed out:

> *"When licensing becomes necessary for gun ownership, government officials will deny to the great majority of Americans the right to have a license, just as city or police officials now do in Denver, New York, Washington, D.C. and dozens of American cities where crime is exploding, where innocent law-abiding citizens cannot buy or own a handgun for protection, but where unlicensed criminals continue to buy, steal, and own firearms at will.*
>
> *"Only good citizens will be licensed (or denied licenses). The criminals will obviously never apply. Licensing will simply disarm the good guys or the victims as it has in New York City - where crime and murder (often with illegal handguns) is exploding.*
>
> *"The BATF will be in charge of licensing gun owners just as they have been in charge of harassing gun dealers in recent years. Do you really want the people who initiated the Weaver/Ruby Creek massacre and shot up the Branch Davidians in Waco in charge of licensing you for firearms, checking out your home for compliance, etc.?*
>
> *"Licensing of American gun owners is unconstitutional and totally flies in the face of everything our founding fathers said about firearms ownership Americans."* Don McAlvany, *McAlvany Intelligence Digest*, December 1993

In the view of the founding fathers and everyone else who knows their history and/or politics, gun registration and then confiscation is necessary to deny people the rest of their fundamental freedoms.

Whether you own a gun or not, if you value your freedom, you had better stand up and be counted before it is too late. Not only should you let your federal and state politicians know that you will fight against any more gun control laws, but tell them that you expect them to honor their oaths of office to protect and defend the Constitution and fight for repeal of the unconstitutional gun control laws already on the books.

Police officers are four times more likely to be killed by their own guns than by assault weapons. Should we take their guns away from them to protect them from themselves? That's rather absurd. However, let me rephrase that slightly. People are far more likely to kill or wound an attacker than they are to injure themselves or an innocent party. Private citizens actually have a much better ratio than law enforcement personnel do. Yet, we should take away their capability and right to defend themselves, right?

Obviously, the object of the gun control laws are to rule over the people, not serve them. According to the FBI, police officers have a greater chance of being killed by a knife than by an assault rifle. Shouldn't we outlaw knives as well? More people are killed by automobiles each year than all kinds of weapons combined. Shouldn't we outlaw cars? There is nothing in the Constitution about any right to have a car as there is about guns. The gun control laws are not about crime. They are about government tyranny.

The old Wild West, where everyone could carry a gun, was not nearly as violent as portrayed in the movies or as our cities are today. The amount of violent crime per capita was tiny compared with what we have today. Violent crime against women was virtually unknown. The few people who engaged in it weren't around very long. Robbing a bank usually took a group of people and some of them usually didn't make it out of town. Today, you can hold up a bank with a written note.

Before Americans can be shoved down the path to the New World Order, our politicians and the CFR/TLC crowd know that they have to disarm us. Before the Communists came into full power in various countries, they had to disarm the people. Before the Nazis came to full power in Germany, they had to disarm the people. Our 1968 Gun Control Act is virtually identical to the first gun control law in Nazi Germany. If you think that confiscation can't happen here, it already has, in New York City. First registration, then confiscation. In Poland, General Jaruzelski canceled all gun licenses. Using the registration lists, he confiscated all of the registered guns. The story was the same in Russia, China, and Cuba and will soon be true in South Africa. Are we next on the list? Former California State Senator Bill Richardson has determined that the California police are monitoring the homes of people who registered their firearms (their names are stored on a special "hazard" database for police) and consider these people to be armed and dangerous. In other words, if you comply with the law and register your gun, you are then considered "armed and dangerous", which is police language for shoot to kill first and ask questions later. Surely, you want to register your gun as soon as possible so you can get on their shoot first list.

Japan, Colombia (of all countries), and a number of other nations are working diligently to disarm the American people. Why? What business is it of theirs? Or are powerful people like those on the CFR/TLC using their influence through these governments?

One of the vehicles for disarming us is a U.N. treaty on gun control that will override the Constitution. Don't let anybody tell you that such a treaty can't do that. The courts have already ruled that the U.N. charter overrides the Constitution and so do any other treaties that Congress approves. This is clearly forbidden in the Constitution, but who cares?

Knowledgeable people who know their history and the insatiable appetite for more power that a little power brings are preparing to defend themselves. The militias that the media and the government are attacking so voraciously did not exist a few years ago. They came into being as a reaction to the government's illegal actions at places like Ruby Ridge and Waco, as well as the obvious ongoing attempt to disarm everyone. They will grow, as the government's attempts at suppression and disarmament grow, in an ever-increasing spiral. The government, in effect, caused the militia movement in the first place, and the only reasonable way to stop it is to stop the clearly illegal acts of the government against its own citizens. Congress, when are you going to do something about the agencies who trample on people's constitutional rights, instead of passing legislation that gives these agencies more power to do so?

The Second Amendment was not about hunting or target practice. Those are assumed and incidental. According to the people who wrote and debated it, it was there for three reasons. They are: 1) To aid in defending the country against a foreign power. 2) To allow a citizen to defend himself, his family, and his property against criminal and hostile individuals. 3) To protect himself, his family and his property against *his own government.* Our founding fathers knew history and human nature. Originally, this amendment was number eight. It was moved up to number two, because they felt it was the second most important right of all those set forth in the Bill of Rights. It is the second most important area of fundamental rights and that is why it is under such relentless attack.

When the assault weapons ban was about to be voted on and voted down, some pro-gun control people showed House Minority Leader Bob Michaels a video of someone shooting down some innocent people with a so-called assault weapon. He changed his usually strong defense of the Constitution with the excuse that no one needed these kinds of weapons to go hunting. That's not the point. The Second Amendment is not about hunting. It is about self-defense. If someone at that shooting scene would have had a concealed weapon, that gunman probably would have killed fewer, if any people. The "disarm the law-abiding citizens" proponents have, over the years, very cleverly redefined the Second Amendment to turn it into a discussion of hunting rather than self defense.

Legislatures have bowed to the political pressure from these well organized and financed groups to continuously nibble away at the rights guaranteed by the Constitution. The courts have also bought into the redefined debate and upheld many blatantly unconstitutional laws that fly in the face of the Second Amendment.

The idea was and is: Second Amendment = Self-defense, from *whomever!*

> *"Guns are not the problem. On the contrary, criminal penalties and laws that disarm honest law-abiding citizens are responsible for giving criminals a safer working environment."…. "The vast majority of gun owners cite protection from crime as one of the main*

reasons they own a gun. And for good reason. Americans use guns for self-protection over 1 million times a year. In 98% of the cases, they simply brandish the weapon or fire a warning shot." - Don McAlvany, *McAlvany Intelligence Digest*, December 1993

Dear Mr. Clinton: If you really want to control crime, rather than control people for your own ends, get all the obviously unconstitutional gun control laws repealed so that people can defend themselves. George Washington and all the others quoted above knew the necessity of a well-armed citizenry, why don't you?

A District of Columbia Court of Appeals has ruled that "the police responsibility is only to the public at large, and not to individual members of the community." In other words, the courts have ruled that the government has no duty to protect individual citizens from crime. This means that you, not the police, the courts, nor anyone else are the ultimate defender of your own life and safety and that of your family. Most sheriffs and local police officers are opposed to disarming the public. They know full well that they can't really protect people. Every criminal shot in self defense is one more that they don't have to risk their lives going after. It is mostly the big city police administrators, who depend on federal money who are backing Washington's drive to disarm anyone who might disagree with the politicians and bureaucrats.

If you think the Republicans are going to defend your constitutional rights, think again. Several times Bob Dole has betrayed the gun owners of America. When a Senate filibuster would have quashed the Feinstein amendment, Dole *quietly* agreed with then Democratic Majority Leader Mitchell not to allow a filibuster. When the Brady Bill was dead in the water and there were only a few senators left around, George Mitchell asked if there were any objections to a compromise worked out behind closed doors by Republican (Dole) and Democratic (Mitchell) leaders anxious to finish one last piece of legislation. There were no objections, of course. So Vice President Al Gore declared that the Brady bill had been passed by unanimous consent of the three remaining senators. Three senators, with the help of Vice President Al Gore, passed the unconstitutional (as declared by several courts) Brady Bill, thanks to Bob Dole. He has not only voted for many laws reducing your constitutional rights, but sponsored some of them such as the Anti-Terrorist bill which is the most blatant attack on constitutional rights yet enacted.

I could go on and on about things like the BATF illegally compiling files of gun owners in clear violation of the law which specifically forbids them from doing so. Then there is the now highly politicized FBI which is blatantly lobbying for gun control according to their own documents in clear violation of the federal law. The FBI's stated goals include:

1. The mandatory licensing of all handgun owners;
2. A ban on the manufacture, transport and possession of semi-automatic firearms, with no compensation for current owners;
3. A ban on high performance ammunition commonly used for hunting and self-defense;
4. A taxpayer-funded handgun buy-back programs with an amnesty period;
5. Passage of the Brady Bill national waiting period;
6. Reallocation of FBI resources to increase regulation of federally licensed firearms dealers.

As for number six above, the BATF already has an all out campaign to get rid of all gun dealers, now the FBI is going to join the campaign. Have you bought any guns you might want yet?

I have spent an inordinate amount of time on the Second Amendment for several reasons. First of all, a well-armed citizenry is the ultimate defense against foreign or domestic tyranny. For that reason, it is probably the main battleground for those who want to strip us of our freedom. Second, it is one of the clearest statements of our rights – rights the government is forbidden to even touch. Yet, these rights already are largely gone. If we can't defend these rights, we can't defend any of our rights. Millions of Americans have died to defend those rights. What are you willing to do while there is still time?

Both the liberal media and Bill Clinton are trying to program us so that whenever we hear anything about crime, terrorists, gangs, or drugs, we think that guns are the problem. These people will do anything to confiscate our guns.

Then what are you supposed to do if someone breaks into your house? Ask him what his intentions are? Maybe you should tell him that guns cause crime, and that since we have no guns anymore, there shouldn't be any more crime, and he should just leave.

While Puerto Rico's gun laws are as strict as any in America, the murder rate is skyrocketing.

The courts are going nuts about people who do use a gun to defend themselves. If you have to shoot someone, you had better make sure you have several witnesses or a videotape of him coming at you with a weapon. If he turns just as you fire and you hit him in the side or the back, you are in real trouble. If you do use a gun, make sure you use your son's one-shot 22 target rifle. You may not hurt him, but maybe *you* won't end up in jail. Heaven help you if you use a so-called assault weapon. The gun may be perfectly legal, but you will find yourself on trial instead of the criminal for using excessive force in *defending your life.*

Amendment IV

The right of the people to be secure in their persons, houses, papers, and effects, against unreasonable searches and seizures, shall not be violated, and no Warrants shall issue, but upon probable cause, supported by Oath or affirmation, and particularly describing the place to be searched, and the persons or things to be seized.

This right has been butchered so often, particularly in recent years, that it is virtually ignored. Now, a Republican Congress has passed legislation expanding warrantless searches which are in direct violation of this amendment. Armed agents invade people's homes, with or without warrants, destroy property, injure or kill people, and take whatever they want, even when the items are not specified in the warrant. When they do bother to get a warrant, it often has no real probable cause, but is likely to be a response to an anonymous tip. They lie to get the warrants as in Waco. There isn't any part of this amendment that isn't violated repeatedly every day by federal agents. You had better hope that you don't ever do anything to attract their attention or aggravate them. Some freedom we have! Don't think that you are safe just because you never do anything to attract attention. People have been killed or injured, and had their property confiscated or heavily damaged when the agents got the wrong house.

Amendment V

No person shall be held to answer for a capital, or otherwise infamous crime, unless on a presentment or indictment of a Grand Jury, except in cases arising in the land or naval forces, or in the Militia, when in actual service in time of War or public danger; nor shall any person be subject for the same offence to be twice put in jeopardy of life or limb; nor shall be compelled in any criminal case to be a witness against himself, nor be deprived of life, liberty, or property, without due process of law; nor shall private property be taken for public use, without just compensation.

...nor shall any person be subject for the same offence to be twice put in jeopardy of life or limb; unless they have done something politically incorrect; then the feds will find some federal law to retry them for the same offense, but under different labels.

nor be deprived of life, unless they are still in the womb, or have purchased too many guns in the opinion of the BATF.

liberty, unless Clinton says they are a terrorist.

or property, without due process of law; unless some federal agency wants their property or some agency breaks their door down on a fishing expedition to try to find something that can be used to put them out of business.

nor shall private property be taken for public use, without just compensation. - until the confiscation laws were passed to allow the taking of people's property on the vague suspicion of some wrong-doing, even just an anonymous tip, to enrich the agency involved.

Amendment IX

The enumeration in the Constitution, of certain rights, shall not be construed to deny or disparage others retained by the people.

In other words, the people, as individuals, are endowed with *all* of the rights. Only the powers and authority expressly granted to the government may be exercised by the government.

Amendment X

The powers not delegated to the United States by the Constitution, nor prohibited by it to the States, are reserved to the States respectively, or to the people.

This amendment has probably been violated more than any other. There are thousands of federal laws and millions of regulations that have nothing to do with any powers granted to the federal government. Many of these are on the books because nobody with the appropriate standing challenged them. Others are in force because the Supreme Court found some excuse to allow them because they were "needed" or useful.

Anyone who can afford to do so should come up with some reason to have the proper standing and challenge in court every law passed by Congress and every regulation issued by some bureaucrat. Maybe then Congress will start paying some attention to whether or not they really do have the right to pass a particular law. Given how squirrely the courts have become, any law just might be thrown out. I think that the Attorney Generals of the states should get together, or do it separately if necessary, and challenge every federal law that even remotely trespasses upon state or individual rights. In the past, state's rights have been improperly used to defend segregation and discrimination, but these have been addressed and we have thrown the baby out with the bath water. It is time to reassert both state's rights and individual rights in compliance with the Constitution. Idaho and Florida have joined Colorado in demanding that the federal government respect the Tenth Amendment. It is going to take a lot more than demands. It is going to take more than law suits, particularly given the Supreme Court's traditional rubber band view of the Constitution. It is going to take the election of people who will honor their oath to uphold and defend the Constitution and not pass laws in areas that are not specifically granted to the federal government.

Amendment XIV

Section 1. All persons born or naturalized in the United States, and subject to the jurisdiction thereof, are citizens of the United States and of the State wherein they reside. No State shall make or enforce any law which shall abridge the privileges or immunities of citizens of the United States; nor shall any State deprive any person of life, liberty, or property, without due process of law; nor deny to any person within its jurisdiction the equal protection of the laws.

This amendment was discussed earlier in chapter eleven. I would ask again, how on earth can an amendment that says you cannot deprive anyone of the right to life, be used to sanction the brutal murder of innocent babies? Only in an Alice-in-Wonderland, politically correct court could anything possibly be so twisted.

Amendment XVI

The Congress shall have power to lay and collect taxes on incomes, from whatever source derived, without apportionment among the several States, and without regard to any census of enumeration.

The story of how the state legislatures were hoodwinked into passing this disaster is a book unto itself. The main issue is that it never had the ratification of three-fourths of the states. Ohio rescinded its ratification before the amendment reached the necessary three-fourths. The federal government was so hungry for unlimited access to people's money that it just declared it ratified anyway, ignoring Ohio's reversal, and started collecting it. The rest, as they say, is history, and will be bankruptcy in the near future.

This amendment allowed the federal government to go around the states and directly into people's wallets. Since we are not likely to elect a President who will order the Justice Department to file a challenge against the validity of the amendment, several states, including Ohio, should challenge it in court.

If that doesn't happen, this amendment should be repealed. Prohibition had more public support and it still was repealed. The income tax is a disaster and *far* more unpopular than prohibition. If we only vote for people who promise to repeal it, it will be repealed. The states, having learned their lesson, want the money to stay home where it belongs.

Amendment XVIII

Section 1. After one year from the ratification of this article the manufacture, sale, or transportation of intoxicating liquors within, the importation thereof into, or the exportation thereof from the United States and all territory subject to the jurisdiction thereof for beverage purposes is hereby prohibited.

This attempt to control the behavior of people by passing laws that a large segment of the people did not agree with was an utter failure. It was repealed by the Twenty-first amendment. People obey laws because they respect them. Apparently, the liberals didn't learn anything from this experience and continue to pour out laws to regulate virtually anything they can think of.

It is time that we fought our way back to constitutional government at the ballot box by only voting for people who will actually follow the Constitution.

Chapter Thirteen

Restoring Constitutional Government

"Liberty has never come from government. Liberty has always come from the subjects of government. The history of liberty is the history of resistance." - Woodrow Wilson

"All governments, of course, are against liberty." - H.L. Mencken

"The spirit of resistance to government is so valuable on occasions that I wish it to be always kept alive." - Thomas Jefferson

"If there is no place for civil disobedience, then the government has been made autonomous and as such has been put in the place of the living God." - Francis Schaeffer

"When people are oppressed, insulted and abused, and can have no other redress, then it becomes our duty as men, with our eyes to God, to fight for our liberties and properties." - Rev. David Jones, 1776

Now we come to the main purpose of this book. All that has gone before is a preamble to the plea I now want to lay before you. Neither the Republicans nor the Democrats are going to turn us away from a dictatorial, one-world government. They get far too much money from those who are pushing us toward that goal and/or believe that it is for the world's good. Either they haven't read their history and don't know very much about human nature, or they have been selling us down the drain deliberately. The Democrats have been like a high-speed passenger train heading down the track towards socialism while the Republicans have been more like a freight train. Both of them are taking us in the wrong direction. The Republican's recent actions haven't even slowed the train down. The various anti-crime type bills have only stripped us of more of our constitutional rights. Neither party is doing anything about the smothering bureaucracies. We need to not only stop the train, but throw it into reverse and begin heading back towards the basic principles embodied in the Constitution and the freedom we once had. The "Republocrats" will never reverse the train.

We must have a new *majority* party. We have lots of third parties, but none of them have a chance of displacing the two major parties. They are identified with a philosophy, a single issue or a person. After Ross Perot's performance in the last election, he cannot possibly be elected. He is right about NAFTA and the need to do something significant about the budget deficit, but he is very wrong about doing it by massive tax raises. This would shatter a failing economy. Campaigning on the idea of raising taxes is going nowhere. We need to reduce taxes significantly and cut government spending drastically. The votes Perot received were primarily protest votes and you can't win with just protest votes. You have to have a basic philosophy that people can easily understand and relate to. A Gallup Poll reports that sixty-two percent of Americans favor a new political party. All that we have to do is get up off the couch and make it happen.

The Libertarian party wants far less government, but they have shown they cannot win anything and many people disagree with some of their philosophy. The Tax Payers' party would probably agree with most of the constitutional principles mentioned in this book, but they are identified with a particular issue and have not communicated their basic philosophy effectively. The single issue name condemns them to a narrow niche. If we all join forces, we can make a difference. If even one of these groups insist on going its own way and running its own candidates, our chance of redirecting our socialistic, people-controlling government is virtually nil. To paraphrase Ben Franklin, we must all hang together, or we will all be jailed/shot/controlled separately.

I am convinced that only a Constitution Party has any chance of achieving massive public support. Only a party that is primarily committed to following the Constitution and its limits on federal government power would have a clearly identified foundation and its very name would convey that commitment. When twenty-five percent of American adults can't find the Pacific Ocean on a world map, the name has to convey the philosophy clearly.

The foundation principle of the Constitution Party should be that the Constitution means what it says and says what it means to say. The Constitution Party's candidates would be committed to vote only for things that are clearly, explicitly allowed the federal government by the Constitution. They would not vote for anything that was not clearly sanctioned by the Constitution, no matter how popular it might be. They would be committed to repeal any laws and regulations that are not clearly within the Constitution's grant of limited power. In other words, they would be among the very few politicians who actually live up to their oath of office to protect and defend the Constitution.

If these candidates were elected and they repealed such laws, they would automatically reduce government by at least sixty percent. Such a party and commitments are our only chance of rolling back the mountain of confused, conflicting regulations that we are buried under and regaining our freedom. Here are some specifics that need to be dealt with.

We must stop the President and the Supreme Court from legislating and usurping the powers of Congress. In the case of the Executive Branch, that means rescinding those extremely dangerous and unconstitutional executive orders that supposedly give the President the power to declare an emergency and virtually suspend the Constitution. We must also stop the various federal agencies from making or enforcing regulations that Congress did not specially pass.

As far as the Supreme Court is concerned, we need to elect a President who will appoint judges that believe the Constitution means what it says and that Congress, as the representative of the people, is the correct constitutional institution for legislating solutions to their problems. According to the Constitution, the executive branch is to carry out the laws and the courts are to apply and enforce them, *not* write them, as both branches often do now. We need judges who understand that the Constitution deliberately put chains on the federal government and that its authors did not intend it to be a rubber band that could stretch anywhere as needed to accommodate the latest social fad.

Social Security and Medicare must be privatized (with federal guarantees) if they are to survive through another decade. Any business that the government is now in would be privatized. Personal income taxes would be abandoned. The wildly proliferating law enforcement agencies would be consolidated into one *investigating* agency. No more squabbling, turf battles and non-cooperation between agencies. One agency would be much more effective and efficient. No more renegade BATF. Law *enforcement* would be the function of the Justice Department in prosecuting cases and the courts in enforcing the law. The mountains of federal law and regulations would be reduced to those that actually have something to do with interstate commerce and other areas that states cannot effectively deal with and are specifically authorized by the Constitution.

The Federal Reserve Act would be repealed, and the control of the money supply would be returned to Congress, where the Constitution requires it to be rather than privately owned banks. Going into debt by the federal government would not be allowed except in declared wars, and a reasonable plan for paying off the national debt would be implemented. An honest money system, backed by gold, would be set up with safeguards to keep the international bankers from manipulating it as they did in the 1920s.

The number of departments in the executive branch would be reduced to six. They would be Defense, Justice, Treasury, State, Commerce and Labor, and Interior. Veterans' affairs would be part of the Defense Department. Commerce and Labor would be combined. Agriculture would go into Commerce and Labor, since it involves both. Agencies like the FTC and the FCC would be in Commerce. Most departments and agencies would be eliminated. The FDA would be abolished. The food part would go to Commerce and Labor (Agriculture) where it belongs. The drug part would go to a different area of the Commerce and Labor Department.

> *"That to compel a man to furnish contributions of money for the propagation of opinions which he disbelieves is sinful and tyrannical."* - Virginia Statute of Religious Freedom, 1786

All subsidies and wealth confiscation/redistribution schemes such as welfare, would be eliminated. There is no constitutional authority for such things. Welfare is the prerogative of the states, communities, and private charities. Congress should pass a law that all existing federal laws and regulations would expire ten years after the enactment. That would mean that Congress would have to spend their time replacing those laws that were truly necessary. Wouldn't that keep Congress from passing very many new laws? Hopefully, yes. They would be too busy trying to decide what was truly essential in the old laws to come up with any new mischief. Congress should be required to pass all regulations. After all, they have the force of law which Congress is elected to enact. Let the government agencies propose the regulations and Congress actually pass those that are needed. Won't that reduce the number of regulations? Hopefully, yes.

Let's review briefly what we have now.

> *"It has been said here many times tonight that we want to make the Senate the same as everyone else, that we want to treat Senators the same as everyone else, that we want to have the Senate treated the same as the private sector. Mr. President, not a single senator wants that."* - George Mitchell.

Clearly, Mr. Mitchell thinks he is part of American's "royalty". The imperial Congress wants to be above the law. That's a large part of the problem. Mr. Mitchell and many like him need to be replaced.

One of Mark Twain's more famous quips was: "Congress is the only criminal class native to America." No one knows whether or not he was joking, but the recent House scandals are just a few of many such incidents throughout our history that makes the activities of congressmen highly suspect. Actually, the petty house hijinks of late are not the real danger to America. Many members of Congress, who would never dream of cashing in campaign contributions for personal gain through the House Post Office, are nevertheless throwing away our sovereignty, production capabilities, and jobs in return for the immense campaign contributions of the multinational banks and corporations.

We have been taught to look to the government to solve our problems. How is a group of people who can't even balance a budget and are mostly lawyers who never had to produce anything, going to solve all our problems? Yet, the liberals/socialists that control our education system keep brainwashing our kids with this same drivel. For instance, one fifth-grade social studies book teaches our kids the following: "Today, when people lose their jobs, they can get some money from the government." Actually, they get it from the "insurance funds" that they and their employees paid for, but in the socialist doublespeak, it comes from the government. It gets worse. Next, the book says: "Today, families who do not have enough money for food can get money from the government". True, but there is no mention of the substantial contribution in this area from private charities. Next, comes: "Today, families who cannot afford to pay their rent can get help from the government." In case you didn't get the message that they are teaching our kids, it is: "Look to the government to solve our social problems and meet everyone's basic need."

Pure socialist propaganda is being pumped into our children at our expense. In case you think that this is an isolated incident, a junior high textbook says: "People (us) were no longer content to live as their forefathers had lived. They wanted richer, fuller lives. They wanted the government to help makes their lives rich and full." Fat chance! In the ever-growing grasp for ever more power by the ever increasing bureaucracy, the false promise that government can improve our lives beyond its basic function of protection, continues to be promoted, and continues to be proven a lie, from the former Soviet Union to America.

This false promise falters on a number of basic principles, including, most of all, human nature. What most politicians seem to forget or ignore is the basic principle that people normally tend to act in their own interests. What people tend to forget is that to run any "solution" through any bureaucracy means that a lot of the money will be siphoned off to support the bureaucracy, and a lot of red tape will be created to slow down and frustrate the process. It is the nature of the beast. Just one example of the government trying to fix things is given in a quote from Ronald Reagan back in 1972:

> *"Forty years ago, government decided to solve the problem of the farmers. Now, forty years and scores of billions later, we only have a third as many farmers as when we started, but we have three times as many employees in the Department of Agriculture."*

Thanks to government programs that give billions of dollars in subsidies to giant agriculture corporations, there has been a massive shift of land ownership from family farms to these large corporations.

From the beginning of government at the tribal level, the basic function of government was protection. People organized for self-defense against common enemies and to establish some form of justice and behavior standards to protect themselves from one another. This is the only proper, necessary, and legitimate function of government. Everything else can be (and should be) carried out by private efforts at far less cost and with greater efficiency, as clearly demonstrated by the post office. The redistribution of wealth (a primary aim of Socialism), which is destroying our economy and is our government's main pastime to keep the incumbents in power, is nothing more than "legalized" theft that is blatantly unconstitutional and unjust.

> *"An excellent example of this leftward swing of the Republican party is the recent invitation by the National Republican Senatorial Committee (dominated by Senator Phil Gramm-R-TX) to have its key donors addressed by Mikhail Gorbachev, former general Secretary of the Communist Party of the Soviet Union. President Gorbachev received $70,000 from the Republicans for his efforts on behalf of raising funds for the GOP.*
>
> *Now think about it! Here is a man who came up through the ranks of the KGB over the bodies of thousands of his opponents; who presided over the extermination of over one million Afghan men, women, and children in the 1980s, who remains a hard-core Marxist/Leninist to this day, and whom the Republicans under George Bush desperately tried to keep in power as his evil communist empire was collapsing around him – here is an evil man whom the Republicans fete, honor, and pay $70,000 to help them raise funds from the Republican faithful.*

"As Mr. Gorbachev said in 1987: 'In October 1917 we parted with the Old World, rejecting it once and for all. We are moving toward a new world, the world of Communism. We shall never turn off that road.' And in a 1989 speech to the Soviet Congress he stated: 'I am a Communist, a convinced Communist; for some that may be a fantasy, but for me, that is my main goal." Don McAlvany, *McAlvany Intelligence Advisor*, December, 1993.

World domination is still Gorbachev's and the Socialists'/ Communists' main goal.

"The more you observe politics, the more you've got to admit that each party is worse than the other." - Will Rogers

Is there any real difference between the Republicans and Democrats when it comes to government control over people? Try this quote:

"They're just a couple of pot-smoking, draft-ducking, scandal-tainted Southern boys who found politics to be a good life without heavy lifting. I'm talking about Newt Gingrich and Bill Clinton-peas in a pod. They're more alike than the media-driven feud would have you believe. Neither can shut up. Both are undisciplined pop-off artists who drive supporters nuts with off-the-cuff blabbing ... Each has shadows. Whitewater's real estate ghosts haunt Clinton. Gingrich had 22 overdrafts in the House banking scandal And the House ethics panel is checking into his complicated GOPAC political fund. In truth, Bill and Newt care far less about money than power. They're blabbers, history buffs, idea wonks. Despite their feud, Bill and Newt are mirrors of each other – small-town, pot-smoking, draft-dodging, scandal-marred political careerists." - Syndicated Columnist Sandy Grady

276 - Restoring America

Here's a classic piece of Orwellian newspeak by Newt Gingrich. He told *Time Magazine*, "I'm for limited government, but a very strong limited government." What on earth does that mean? Here is an insightful quote about Newt's approach to welfare:

> *"I am in almost full agreement with Newt Gingrich on the issue of welfare...I'm for cutting welfare too, but only if we begin at the top where the greatest offenders live – like, say, members of the U.S. Congress.*
>
> *"They waste more money in a month than most alleged welfare chiselers do in a year. Many of our so-called leaders enjoy benefits that few Americans have ... Members of Congress earn a base pay of $133,644 yearly. Yet, many continue to enjoy such perks as low-cost haircuts, free airport parking, free car washes, use of limousines and government-paid foreign travel.. Representative Bob Michael (R-IL) was cited for spending $58,070 in campaign funds on golf-related expenses... Welfare reform is long overdue. But let it begin in Washington with the Gingriches and the Bob Doles, where it can best be a afforded."* - Claude Lewis, Knight-Ridder Newspapers.

Here is a quote about all those congressional hearings that you see going on, by a Congressman no less.

> *"We create the Government that shafts you, and then you're supposed to thank us for protecting you from it."*
> - Rep. Vin Weber

Most congressional hearings are nothing more than grandstanding before the cameras. Just watch and see what comes out of the hearings on Waco, Ruby Ridge, and all the other crimes and the ignoring of basic rights by the various government agencies. There may be a lot of noise and smoke, but very little will change. A number of years ago, Congress held hearings on the BATF's abuses, and branded them a renegade agency, and then did nothing.

Congress is just as responsible for Waco, Ruby Ridge, and all the rest of the abuses as the FBI, the BATF, the Justice Department, and the Treasury Department are. They could have reined in these agencies years ago. Instead, they recently gave them expanded powers and more money to do more of the same.

Generating regulations is sloughed off to bureaucrats. Congress can be very specific when they want credit for some pork-barrel legislation. They only avoid specifics when they want to be able to blame the government agencies who come up with unpopular regulations they ducked making. Here's one such detailed regulation from Congress: "Not later than 30 days after the date of enactment of this Act, in Chambersburg, Pennsylvania, at both the intersection of Lincoln Way and Sixth Street and the intersection of Lincoln Way and Coldbrook Avenue, the Pennsylvania Department of Transportation shall include an exclusive pedestrian phase in the existing sequence between the hours of 2:45 and 3:45 p.m. on weekdays."

Not only does this prove that Congress has the time to be extremely specific about enacting laws, regulations, or whatever you want to call them, it also shows just how far Congress has gone in violating the Tenth Amendment. What business is it of Congress to tell the City of Chambersburg how to sequence their traffic signals? Congress has been granted absolutely no authority to do so by the Constitution, but it ignores such minor details.

By abandoning their job of legislating specific details in the laws, Congress attempts to avoid being held responsible. Instead, they pass vague dumpsters for bureaucrats to dump tons of regulations into for American business and individuals to carry around on their backs. Of course, they exempt themselves from obeying any of these, per Mitchell's quote above. The result of this is that people find themselves hauled into court for some claimed infraction, and even the agency hauling them into court doesn't know what the law is until a judge hands down his ruling. The FDA is particularly infamous for this. If the citizens complain loud enough, then Congress holds hearings and occasionally blames the agency for going too far. It's a great system they have for passing the buck.

Term limits are a bad idea whose time has come. It is the lesser of two evils, and it is the only hope we have of returning to the concept of citizen legislators. Maybe we can get some engineers, farmers, small business owners, and others who have to live in the real world to replace those masses of lawyers that now dominate Congress and state legislatures.

Can you believe a politician? Lest you think getting elected isn't more important than any principle or scruples, read this.

> *"I have to be – you heard of a whore? I'm a whore. I am a political whore. And I'm going to play it to the hilt....Shaking hands and kissing everybody. I mean I'm here to get elected. I'll be going to a lot of funeral homes. Just walk in and-if I faintly remember who those people are-just walk in and shed a little tear and sign my name and take off."* - Rep. Joe Kolter.

When getting reelected (forever) is the only thing that politicians care about, we are going to have rampant corruption and special interest groups controlling our lives. It will take a tremendous revolution in virtually every area of politics, government, taxes, education and moral values to get this country back on the right track. For that to happen, there will have to be a significant revolution in people's thinking. For three generations, we have been programmed to look to the federal government for solutions to many problems. We must have a revolution in people's attitudes toward and understanding of government.

It is time for a real revolution. Not a revolution with bullets and bombs, but one through ballot boxes. Returning to true constitutional government is impossible with either the Democrats or Republicans. For the Constitution Party to have a chance to keep us out of a one-world government or our own government from becoming a socialist monster, every other small party will have to join in the struggle by supporting the Constitution Party. They can do nothing on their own, except perhaps give the election to Clinton, which would be a disaster. He has repeatedly shown that he wants to rule by decree (executive order), and he is perfectly capable of suspending the Constitution, particularly if he is in his second term.

In my opinion, Bill Clinton is the most dangerous man to ever inhabit the White House. He has left a trail of thirty bodies along his trail to the governor's chair and the White House. (See the book or video, *The Clinton Chronicles*.) Whether he had any direct knowledge or involvement in these events, we will probably never know, but quite a few "friends of Bill" were involved in these situations. Other people were severely beaten. Some of these incidents involved recovering files, pictures and video tapes of Clinton. Was Vince Foster murdered because he knew too much and was cracking?

> *"The coverage of the Foster story suggests that leading newspapers in America are helping to disguise the probable murder of a high-ranking official of the U.S. Government....Almost every detail that is leaked to the service reinforces that he was murdered and that his murder was staged to look like a suicide."* - Jame Dale Davidson, *Strategic Investment.*

With the Presidential executive orders already law, plus the anti-terrorist laws, Clinton can wipe out his enemies or completely take over the country as a dictator with a stroke of the pen. These laws must be repealed.

We need to build a major party from scratch in less than a year. The window of opportunity does not extend to the election of 2000. By then, it will be too late. If you don't understand that, I believe that you do not clearly understand where we are in history and the coming together of so many physical and historical cycles. Rockefeller knows that this window is very small for his forces to do away with our Constitution and impose theirs. We had better understand that as well.

We need candidates at every level who are committed to constitutional government. We need school board candidates that believe in local control of schools. We need precinct committeemen and women in every precinct. We need workers who will pass out literature and discuss issues with people. Lastly, we need money, a lot of it. That money will not come from PACs, big corporations, or any other special interest groups. It has to come from us; you and I. We will lose much more than the a few dollars if we don't elect people who will protect us from our own government. Ask the survivors of Waco.

> *"All that is necessary for the triumph of evil is that good men do nothing."* - Edmund Burke

> *"I am only one, but I am one. I cannot do everything, but I can do something. What I can do, I should do, and, with the help of God, I will do."* - Everette Hale

Even if you never ran for office or participated in politics in any way, you need to get involved. Even if you seldom vote, you need to get involved, or you might lose the chance to vote. Millions of Americans have died to preserve our constitutional government. Your freedom, independence, and financial assets are very much at risk now. What are you willing to do?

Chapter Fourteen

Survival Manual

"Debtors are always backward-looking, not forward looking. Debtors always turn their attention to paying for the depreciating asset they already possess rather than saving for what they might own outright in the future. Debtors are servants rather than masters. Debtors have no real vision of the future because their economic situation is governed by the past." - Dr. Andrew Sandlin, *Chalcedon Report*.

Restoring constitutional government to America will not happen overnight. Before that can be accomplished, we will go through some very difficult times. When the economy collapses, many people will be clamoring for the government to do something to create jobs, pay unemployment benefits, keep their homes from being repossessed, and control the riots and rampant crime that will be destroying the cities. They will gladly give up their constitutional rights and freedom for food and stability. It will be very difficult to convince people that true freedom and minimum government interference will bring a much more rapid recovery without the government fouling things up, as they did during the Depression. If the Constitutional Party could elect a President and at least one-third of Congress in 1996, there is a chance that the economic crash could be softened considerably, but not eliminated. We are too deeply in debt.

If you are wondering why I have included a survival section in this book, it is because some of us have to come out of the mess with enough assets to fuel the recovery. Also, there is no use losing what you have earned just because the politicians are far more interested in keeping their jobs than in keeping the country solvent.

282 - Restoring America

GET OUT OF DEBT

The borrower is slave to the lender. If you don't believe that, try not making your house payments for a few months. Americans have mortgaged their future like no people in history. There is no way the government, corporate, and private debt can ever be repaid in real money. Many people are going to lose everything they have worked for all their lives. Don't be one of them. Our real income and standard of living have been declining for twenty years, and yet we have been borrowing more and more.

We are entering a time of worldwide deflation. Japan is leading the way, with a real estate market and stock market that have lost half their values. Most of their banks are on the brink of collapse, with over a trillion dollars in bad loans. The Japanese politicians want to pull back their assets and let the U.S. drown in red ink. However, the bankers know that if that happens, they will sink also. They are inflating their money just like we do for every Presidential election to try to keep the economy growing. With much of that phony money, they are buying gold by the ton. With the rest of it, they are buying dollars. The real reason that the dollar is presently gaining on the yen when it is in worse shape is that the Japanese banks are "printing" the money to buy dollars faster than we are currently "printing" ours, to help their exports which are dropping dangerously, and unemployment is spreading. Money, real money, is going to become very scarce. Get out of debt as fast as you can to protect your real assets, like your home.

Many people will say, "How can I do that? I can barely keep my head above water now." Let me give you a few ideas. First of all, make out a budget, a real one that reflects what you are actually spending. It will probably be a brutal shock. At the least, it is no fun, but we are talking about your financial survival. The lifestyle you have been used to is about to come to an end. Either prepare for it and save what you can or perhaps lose it all. Take a hard look at your budget. What can you cut out and do without? Cut it out. Then, pretend that you lost your job. You very well may. Now, what would you cut out? Do it. How much money would you have left each month? Take that money and start paying off your debts, credit cards first if you carry a balance. Pay off the highest interest rate debts first. If you can only work on one or two, pick the smallest ones. When those are paid off, apply those payments to the next smallest. Don't let up. People who have felt they were hopelessly in debt have cleared up all their debts in two to three years. You may not have that much time, but do the best you can.

Change your buying habits. Buy cheaper fresh foods, rather than processed ones. You will pay less and be much healthier. Make a shopping list and stick to it. Avoid impulse buying. Shop for things that you really need, such as clothes, at garage sales. If that offends your pride, you are going to be in real trouble. Sell anything that you really don't need and apply the money to your debts while other people still have the money to buy your items. If all that sounds too desperate, don't say you weren't warned.

After paying off any debts you might have, it is time to plan on some defensive investments. Don't go near the stock or bond markets. They will soon crash. You want real assets, not some worthless paper that represents someone else's debts. First of all, buy some "junk" silver coins in bulk. They can presently be bought for virtually their melt value. The time is coming when only real money, silver or gold, will be accepted as a medium of exchange. Then, begin buying common gold coins which can be bought at two to three percent over their bullion value. Gold coins will be needed for larger purchases and can be easily hidden if necessary. There is a very real possibility of the government confiscating gold again as their need for money becomes desperate. Whether you comply is a decision you will have to make if that happens. Don't buy the story that the government won't confiscate numistic or semi-numistic coins so you should pay a premium for those. If things become as bad as I believe they will, those niceties won't matter. You want these coins for survival in a shattered economy.

It is true that numistic coins may well outperform bullion coins when the government starts printing money like mad to pay its bills and inflation goes through the roof, but that won't last long. It will be just the last stage before we are forced back to an honest money system. Investing in rare coins requires free cash, and that will be in short supply, so eventually there will be few buyers. Buying numistic coins can be a good investment for a knowledgeable individual at certain times in the market cycle. You probably aren't one of them, and unscrupulous dealers abound. Know your seller. Don't buy anything from a telephone salesman, no matter what they promise. Remember, you are buying coins at retail and would have to sell them at wholesale. It takes a very large move in the price of gold for you to break even. Your coins are for emergencies, not investment. Under no circumstances should you leave your coins or bullion with anyone else. Get them immediately into your possession.

Stay away from real estate. That used to be the easy way to become a millionaire for average people. No more. Real estate values are going to crash also, for several reasons. Commercial real estate is especially vulnerable in the coming economic chaos, but will decline even without any collapse. In the Information Age, with more people working at home, office buildings could become white elephants when it comes to real estate investment.

Beyond hard cash, you may want to stockpile some non-perishable food. Clinton is already making noises about people hoarding things. If he is President when the bubble bursts, you can be sure he will use the unconstitutional powers in those executive orders to try to make sure you carry your share of the pain from the politicians' stupidity. Also, make sure you have some guns and plenty of ammunition. The further you are from a big city, the less chance you have of needing to use them, but anyone is vulnerable. When people become desperate for food, the thin veneer of civilization vanishes. It wouldn't hurt to learn a thing or two about gardening. There are a number of good books available on becoming self-sufficient. Take the time to read some of them now, before you need the information.

ALTERNATIVE INCOME

As mentioned in the chapter on the economy, try to develop some skill that people will need as a second source of income now and perhaps your primary income when the economy crashes or when you retire. If you do lose your job, it will help you survive. Find a way to build a second income, preferably from home. Personal computers are making that much easier to do. The number of different kinds of opportunities is exploding.

> *"A new information class is being created which will be the dominant class of the next century, just as much as the feudal barons were in the 12th century or the newly empowered bourgeoisie in the 19th. Those who can profit by the Information Age will be those who have the financial capital or the personal skills to do so. "* -
> Lord Rees-Mogg, *Strategic Investment*

There are some very good books on building a second income. Get some books from your local library. There is a lot of junk being peddled and the library will tend to have the better books. Stay away from any form of sales. Competition from sales experts will become intense for the few spendable dollars available.

MAINTAINING YOUR HEALTH

One of the most important things you can do is to maintain your general health level. Environmental poisons, poor diets, resistant strains of bacteria, and new lethal viruses will become more and more devastating in the years to come. If you don't have your health, little else matters.

As the allopathic medical monopoly tries to limit your health care choices, largely through the police state tactics of the FDA as well as HMO restrictions, you will need to become much more knowledgeable about taking care of your own health. Besides that, you may not be able to afford the skyrocketing costs which may turn out to be a blessing in disguise. If you understand the intent of Hillary's communistic health care (people control) plan, it should warn you of what is coming. All those fines and jail terms for going to a doctor of your choice were not in her plan to improve your health. It was an all-out attack on alternative and complementary medicine.

Nothing in this section should be taken as a recommendation for self treatment or medication. Any diagnosis, treatment or medication should be provided by a licensed physician. The information is presented solely to alert you to what is available and, in some instances, largely unknown to "conventional" medical doctors. Yet, virtually all of it has been taken originally from medical journals and clinical studies. If you are not getting satisfactory results from conventional doctors, you may want to find a doctor who practices complementary medicine. You may be able to obtain the names of some of them from the Atkins Center in New York City or the American College of Advancement in Medicine which are listed in the Appendix. For more information on alternative medicine, you can purchase *Alternative Medicine – A Definitive Guide*, listed in the Appendix. An excellent synopsis of various forms of alternative medicine and treatments, along with a long list of books on various medical subjects, can be obtained by ordering a copy of the August 1995 issue of the *McAlvany Intelligence Advisor* at P.O. Box 84904, Phoenix, AZ 85071.

Eating as much fresh, unprocessed food as possible is a good place to start. Most fruits and vegetables should be eaten raw or lightly steamed. Beyond that, strong supplementation is needed today. Don't believe the nonsense that you can get all you need from food. Nobody eats enough of the right foods to start with. Much of the food we eat is deficient in nutrients from extensive agriculture. Our immune systems are under continuous attack from environmental poisons and the stress of modern life. The following daily "formula" (or one close to it) for sufficient supplementation and optimum health is recommended by a large number of complementary and alternate physicians and dietary experts.

Vit. A, 5000 IU. Beta Carotene, 25000 IU. Vit. B1 200 mg. Vit. B2 50 mg. Vit. B3 100 mg. Vit. B5 600 mg. Vit. B6 175 mg. Vit. B12 100 mg. Vit. C 3 grams. Biotin 200 mcg. PABA 50 mg. Folic Acid 800 mcg. Bromelain 15 mg. Vit. E 800 IU. Selenium 100 mcg. Zinc 35 mg. Magnesium 1000 mg. Calcium 1000 mg. Potassium 100 mg. Vit. D 200 mg. Chromium Picolinate 50 mcg. Molybdenum 100 mcg. Manganese 5 mg. Iodine 10 mcg. Several sources of quality vitamins with this approximate formula are listed in the Appendix.

To insure the vitamins are properly activated, you should eat plenty of fresh fruit and vegetables with them. At least get your vitamins from a company that manufactures them from food and plant sources that preserve the phytochemicals.

Reducing the amount of refined sugar, corn syrup, and similar sugar products to no more than a pound a year would improve almost everyone's health. Better yet, avoid them all together if possible. However, do not substitute aspartame (Nutra-sweet, Equal, etc.) for sugar. The list of health problems from aspetame fills pages. It has generated far more complaints to the FDA than any other food additive and can cause very serious health problems. You can get the long list of symptoms and problems from the FDA or from the Internet newsgroup misc.health.alternative. One of the components of Aspertame is wood alcohol, which is a deadly poison, yet the vast majority of diet soft drinks contain Aspartame.

AIDS

With the cut-and-drug mentality of conventional medicine, it is no wonder that they are having problems finding a drug to kill the various HIV viruses. Until (if) they do, what they should be doing is strengthening the immune system to deal with the virus as it was designed to do. There are ways to kill or weaken the virus that let a strengthened immune system conquer it. The following items can be use to either strengthen the immune system or attack the virus or both:

Vitamin A. Beta Carotene. Vitamin C. Vitamin E. Bioflavonoids. Selenium. Zinc. Garlic. Maitake mushrooms. KH3. Flaxseed oil. Licorice root. Echinacea. Astragalus. Goldenseal. Green Tea. Aloe Vera. Chlorella. Spirulina. DHEA. Thymus gland extract. Melatonin. Ukrain. N-Acetyl-L-cysteine. Glutathione.

There are three proven ways to attack the virus itself, beyond what garlic and some of the other compounds can do. One is bio-oxidation. This consists of injection or infusion of either hydrogen peroxide or ozone into the body. Excess oxygen kills viruses. A few doctors in this country are using hydrogen peroxide quite successfully. In Germany, over 300 people who were given ozone treatments are now walking around without a trace of HIV. Another successful treatment is photo-luminescence, where the blood is treated with ultraviolet light. This was very effective against both bacterial and viral infections before antibiotics were discovered.

Because of their convenience, the antibiotics swept this technique away. Unfortunately, antibiotics usually have little effect on viruses. This technique is one of the most effective treatments for septicemia. Yet, people die each year from it, because the technique and equipment is only available from a few complimentary physicians. It has proven to be highly effective against the AIDS virus.

Another technique that has been shown to be very effective is to pass minute DC currents through the blood. Nobody knows why this works. Apparently, it disrupts the virus somehow and stimulates the immune system to kill them off. The technique has been proven in laboratory tests and is patented by the Albert Einstein College of Medicine. It is not a do it yourself project, as you could get killed playing around with electricity.

If you want all the specific details on this approach to dealing with HIV, or you just want to know how to build up your own immune system, you can order the booklet *AIDS: A Broader Approach To Treatment*, with the order blank in the back of this book.

ALLERGIES

Most people have at least some allergies. They may not know it, but allergies can cause a bewildering number of symptoms often attributed to something else. The best defense against allergies is the supplementation program at the start of this section. Many household allergies, like dust mites, can be helped by a high filtration air filter. In addition, electronic air filters can remove many viruses and bacteria.

Milk product allergies are more widespread than many people think. Elimination of all milk product from the diet for allergies that are not obviously from pollen, dust mites or similar sources may bring relief. If you suspect any kind of food allergy, the elimination diet is an excellent place to start. Start by eating only watermelon for a couple of days. Slowly add things like cantaloupe and bananas for a few more days. Drink only distilled water. If, at the end of a week, the allergy symptoms are gone or significantly less, you have a food allergy. Carefully add one food at a time until the symptoms return. You can easily identify what is causing the problem. Sometimes it is not a food, but a food additive, so watch for those as well.

Many people have found relief from allergies by simply drinking eight glasses of pure water a day. See the water section below.

ALZHEIMER'S

This devastating disease is skyrocketing for two main reasons. One is that people are living longer. The other is the poisoning of our brains by toxic metals. We now have "normal" lead levels in our brains that are 1000 times what our grandfathers had. Mercury fillings, which are being banned in many European countries because of the overwhelming evidence of their effects on health, are a serious source of various kinds of brain damage, and not just in Alzheimer's. Alzheimer's patients also show a huge amount of aluminum in the brain. There seems to be a genetic factor at work here, but some of it may also be a lifestyle that includes significant aluminum sources. Until more is known in this area, we all should avoid ingesting aluminum into our bodies. The largest source of aluminum is antacids, which millions of Americans consume by the gallon each year. Don't! There are other over-the-counter medications that also contain aluminum. Don't use them. Some deodorants contain aluminum salts that can be absorbed through the skin. Aluminum is used as a food additive. Check the labels. Perhaps worst of all, aluminum is used as a clarifying agent in many municipal water systems. You probably shouldn't be drinking tap water anyway, but this is just one more reason. Other sources of aluminum that should be avoided are pop cans and cooking utensils.

There are some drugs that can slow the destruction of the brain cells and/or improve cognitive abilities to help Alzheimer's patients. Typically, the FDA has tried to stop people from obtaining some of them although they clearly help people. This led to the forming of the Alzheimer's Buyers Club operating out of Switzerland so people could continue to obtain the drugs they needed to help their family members. Some of the agents that help Alzheimer's patients include L-Deprenyl, Acetyl-L-Carnitine and THA combined with Lecithin.

No drug will completely reverse the damage done by these toxic metals. However, there is a perfectly legal procedure that will remove these metals, prevent further damage and allow many of the brain cells to be repaired. It is called chelation. A synthetic protein, EDTA, is slowly injected intravenously into the blood stream. This is a standard legal treatment for lead, mercury, and other heavy metal poisoning. The results of this treatment, coupled with some brain regeneration chemicals and drugs, has been amazing. People who needed help feeding and clothing themselves have been able to return to work. The sooner treatment is started, the better the results. Obviously, if much of the brain has been damaged, the results will be less. However, some improvement has been shown in all cases. If you know someone who has Alzheimer's, have their family get a book entitled *Toxic Metal Syndrome* by Drs. H. Richard Casdorph and Morton Walker. These men are experts in this field. Tell them to give the book to the attending physician. If he won't provide the treatment, get a list of doctors belonging to the American College for Advancement in Medicine, who will. Call 800-532-3688 for information on chelation and a list of members.

ARTHRITIS

Recent studies have shown that chicken collagen offers significant help to rheumatoid arthritis sufferers. Real licorice (glycyrrhizin) is also helpful in many cases. One study reported in the Annals of Internal Medicine stated that arthritis sufferers who took fish oil supplements had significantly less pain and were able to perform much longer. This confirms what other studies have reported. Most complementary physicians are far more effective than conventional doctors in dealing with arthritis.

ASTHMA

In one study, asthma sufferers who took 1000 mg. of vitamin C each day had four times better breathing ability than those who took none. In another study, when they took 1,000 mg. of vitamin C each day, they had less than one-fourth as many attacks. Other asthma sufferers have had their problems nearly or completely eliminated by drinking eight full glasses of pure water a day. See the water section below. The high efficiency air filters mentioned above can help.

CANCER

Cancer is the second leading killer disease, next to heart disease. Again, conventional cut and drug medicine is overlooking the best weapon we have against cancer in their hunt for magic bullets or lucrative treatment procedures. The immune system is designed to destroy defective (cancer) cells. Any cancer patient should do everything possible to build up his or her immune system. Yet, many conventional approaches virtually destroy the immune system.

One of the most effective treatments for many forms of cancer is extremely simple and inexpensive. It was developed by Dr. Johanna Budwig in Europe, who has been nominated for seven Nobel prizes. The main ingredient in this treatment is flaxseed oil. When taken with low-fat cottage cheese, which provides the needed sulphonated proteins, it is a very powerful immune system builder and is very effective against many forms of cancer. She has successfully treated many patients, who have had their immune systems virtually destroyed by radiation and chemotherapy and have been sent home to die. Use only cold-pressed, carefully preserved flaxseed oil, as it oxidizes very easily. Some other agents that are particularly effective in helping the immune system deal with cancer are Maitake mushrooms, Hemogobulous, green tea, garlic, vitamin D, beta carotene and broccoli. "Broccoli contains the most powerful anti-cancer compound ever detected." - *Alternative Medicine Digest No. 4*. Drugs that help fight cancer include Interlukin-2, Alpha Interferon, Levamisole, Tagamet, Isoprinosine, KH3, and Melatonin. To inhibit tumor growth, shark cartilage in large doses reduces the blood supply to the tumors.

Any cancer patient should eat lots of fresh, raw food. Isolated vitamins and chemicals do not have the same effect as the whole combination of nutrients that are available in fresh, raw foods. In fact, if people ate five servings of fruit and vegetables daily and ninety percent of their diet was from whole foods, America's health costs would be cut in half. Don't look for the government to recommend this. It would reduce their power over people and make Social Security even more insolvent.

Breast cancer is a leading cause of female deaths. Besides the general items above, melatonin, which the body produces naturally, is proving to be an effective preventive. The natural production of melatonin in your body decreases with age. Supplementation can be a significant help in prevention of a number of problems., including breast cancer. Melatonin is very safe and very effective for sleep disturbances as it is the natural sleep regulator the body needs. Typically, the FDA is doing everything they can to intimidate the manufacturers and importers into removing it from the market to protect the profits of the drug companies who peddle sleeping pills which are much less effective and cause more problems.

Lignins from flaxseed and Isoflavones from soybeans have shown significant results in preventing and treating both breast cancer and prostate cancer.

One of the most promising developments in the treatment of cancer is a substance called Ukrain. It comes from a plant called celedine. It has such an attraction for cancer cells that it can be used to tell whether a growth is malignant. In preliminary tests, it killed every one of the 60 different tumor cells that the National Cancer Institute uses to test cancer medications. It has been shown to shrink tumors, destroy cancer cells and boost the immune system. It can even help doctors locate tumors and cancer cells. Somehow it seeks out cancer cells. When exposed to ultraviolet, it glows. It is virtually non-toxic, unlike more common cancer treatments. Ask the FDA why studies are not being done in the U.S.

CANKER SORES

These lesions in the mouth are caused by common bacteria. A cut, nick, or bite can provide a place for the bacteria to thrive. Stress, which reduces the immune capability, often helps trigger the infection. The most effective way to treat them is to use a product called ORA5. It is a mixture of alcohol, copper sulfate, iodine and potassium iodine. You can obtain it over the counter at most drug stores. Frequent canker sores indicate a high stress level and/or the need to improve nutrition with raw foods and supplements.

CHRONIC FATIGUE

Virtually everyone complains of fatigue these days. Some of it is from the hurried stress filled lives we lead and its effect on our immune systems. Some of it is caused by environmental contaminants. Viral problems like those caused by the Epstein-Barr virus can be devastating. Sleep disturbances can cause chronic fatigue and fibrositis or fibromyalgia-type problems. The wide variation in causes and symptoms make this problem particularly difficult. It takes a very knowledgeable doctor to track down and eliminate various possibilities. For some people with serious debilitating symptoms, taking guefesson can make a big difference. Typically, they feel much worse at first, but eventually get a great deal of relief. Where viruses are involved, the electrical currents discussed under AIDS above have completely eliminated the symptoms for some people. Licorice (glycurrhizin) can be very effective against CFS. It inhibits cytomegalovirus, one of the main suspects in chronic fatigue. It also stimulates the production of interferon. If that wasn't enough, it rejuvenates the adrenal glands which are almost always exhausted in cases of chronic fatigue. Unfortunately, at the dosages needed, some side effects may appear and it must be taken under the care of a knowledgeable physician.

Fibrositis-type symptoms are typically treated with drugs like amitriptyline or cyclobenzephrine for sleep enhancement and muscle relaxation. These can have some rather unpleasant side effects. If sleep problems are a factor, melatonin may be a much better answer. Again, complementary doctors typically have much better results in this are as they tend to take a much more holistic approach looking at a wide range of factors rather than just trying to alleviate a symptom or two.

COLD SORES

These blisters are usually caused by the herpes simplex 1 virus which almost everyone carries. Frequent blisters may indicate a lack of sufficient calcium in the diet. Supplemental calcium will reduce the frequency. The virus usually stays dormant in the ganglia of the nerves. When stress or other things trigger it, it moves down the nerve to the skin to cause the blister to start developing. At the first sign of tingling or burning, apply ice to the spot every ten minutes for an hour. This helps prevent the virus from moving to the skin. Start taking L-lysine, a common amino acid, at 2,000 to 3,000 mg. a day. L-lysine inhibits virus reproduction.

If you are prone to these sores, as I am, try taking 500 mg. of L-lysine a day. Triple the dose if one starts to develop. If you still get one, rub it with something abrasive to break it open so it can dry out. Apply Melissa Officinalis (herb) to the sore. It will cut the healing time in half. Some health food stores carry a salve or gel that contains this herb, aloe vera, and other agents that speed healing.

COLDS/FLU

For years, the medical establishment denounced Linus Pauling and his mega-vitamin C approach to preventing colds. Well, he lived to be ninety three years old and didn't catch colds or much else. Conventional medicine is now, grudgingly, beginning to acknowledge the effectiveness of vitamin C and other herbs and supplements. If you are on the supplementation at the beginning of this section, you will probably have fewer problems with colds, allergies, or flu. If you do start to get a cold or flu, start taking 500 mg. of Vitamin C every hour at the first sign of symptoms. It is quickly eliminated from the body, and you need a constant supply of it. Also take a herb called Echinacea. If the product contains Astragalus and Reishi, so much the better. These provide a quick, if temporary, boost to the immune system. Triple your intake of garlic, taking it every few hours as it also is quickly eliminated. It not only boosts your immune system, but attacks bacteria and viruses directly. Drink lots of water and rest as much as possible. Hemogobulous will significantly reduce the symptoms and speed your recovery. Double the recommended intake of beta-carotene. Zinc gluconate lozenges taken every two hours can cut the duration of a common cold by seven days. Don't overdo the zinc, as too much can start depressing the immune system.

If you start throwing up, try nibbling on plain dry toast and sipping unsweetened hot tea. It doesn't sound very appetizing, but it works.

DIABETES

Adult diabetes is aggravated by the immense amount of sugar many people eat. Constantly dealing with all that sugar exhausts the pancreas from producing insulin to try to deal with that sugar. Whether you have diabetes or not, virtually eliminating refined sugars from your diet will make you much healthier. Taking chromium picolinate will help your body deal with sugar properly. It can reduce or eliminate some of the medications or insulin that diabetics need. It will help prevent you from exhausting your pancreas.

HEART DISEASE/STROKES

The leading killer is heart disease. Amazingly enough, the clogged arteries that cause many heart attacks and strokes were virtually unknown until well into this century. Despite the fact that a largely rural population ate great quantities of animal fat, milk, butter, and eggs, this problem virtually didn't exist. It wasn't even taught in medical schools. What has changed?

There are four main factors in today's explosion of clogged arteries. One is exercise, or the lack of it. The others are less known and virtually ignored by conventional medicine. Angioplasty and bypass surgery are just too lucrative. They are also dangerous. Hundreds of patients die from these procedures each year, and bypass surgery often results in some brain damage, sometimes, severe.

One of the biggest causes of today's clogged arteries is trans-fatty acids, or hydrogenated vegetable oils. These artificial fats cannot be handled properly by the body and end up being stored in the arteries. These oils are everywhere, starting with margarine. They are included in most processed foods. It is no wonder that virtually every male over fifty and every female over sixty has some plaque in their arteries. This stuff should be outlawed. In five years, heart disease and strokes from clogged arteries would probably be cut in half.

There are two other factors that assist the hydrogenated oils to collect in the arteries. Both are dietary substances, whose uses have also skyrocketed. Refined sugar plays a significant role in irritating the lining of the arteries. Today, people eat a tremendous amount of white sugar, corn syrup (soft drinks), or other forms of sugar. Sugar is everywhere. It should not be allowed in baby food, as that predisposes the child to a lifelong addiction to sugar. The arteries respond to this irritation by trying to protect themselves, coating the irritated area with plaque. Cholesterol, fat and platelets are drawn to the site, and you have atherosclerosis. There are other sources of irritation to the arteries, such as cigarette smoke, high blood pressure, and high cholesterol. Catheters used for angiograms are a major source of irritation to start the process in various places. According to medical researchers, 900,000 unnecessary angiograms are performed each year and 4,500 people die needlessly. Those that don't die have significant injury to the arteries that start the atherosclerosis process at the site of those injuries.

The third agent that helps clog arteries is excess iron in the blood. The iron is necessary to oxidize the fats. The FDA *requires* many foods to be fortified with iron. Other products contain iron to be sold as "fortified". Virtually all multiple vitamins contain iron. While younger women may not get enough iron, men, and women past menopause, should not take iron supplements unless there is evidence of iron-deficiency.

The use of all three of these compounds has dramatically increased, particularly since the 1940s. So have heart disease and strokes from clogged arteries. This is not a coincidence. It is cause and effect. Arteries can be cleared of this goop over a long period of time by the rather severe Pritiken diet. A much quicker way is chelation therapy. Typically, about twenty treatments have most arteries looking like new. Well, maybe not new, but clear of plaque. This approach is far safer than surgery. It costs much less. It is much more effective. It cleans out *all* the arteries, not just those around the heart or in the neck. Again, this treatment is provided by most of the physicians who are members of the American College for Advancement in Medicine as well as a number of other complementary physicians.

Eliminating hydrogenated vegetable oils, cutting your sugar intake at least fifty percent, and eliminating iron supplementation for males and post-menopausal women (unless needed) will virtually guarantee freedom from heart attacks and strokes caused by clogged arteries. There are some other things that you can do. The people of Crete eat a lot of fat in their meals, yet they have one of the lowest heart disease rates in the world. Several studies have traced the results to their high intake of virgin olive oil, which represents about thirty-five percent of their fat intake. Red wine also has been shown to have a beneficial effect on heart disease.

Fortunately, it is now been shown that the beneficial effects of wine can be provided by fresh grape juice without the damage and problems of alcohol.

> *"A review of societies eating a primarily vegetarian diet – no eggs, no cheese, no milk, no meat or the fat of meat - are decidedly unhealthy. The only thing that holds the vegetarian argument together is the lipid hypothesis of atherosclerosis and that turned out to be entirely specious in that they were blaming the wrong fats: animal rather than vegetable, i.e., butter rather than the real culprit - margarine and other vegetable-oil based foods."* - Dr. William Campbell Douglas, *Second Opinion*, Dec. 1994.

> *"Only one 'general type' of fat should be avoided: the artificial, made by industry type of fats that have not evolved with (or haven't been created simultaneously with) the evolution or creation of the human body....Recent research shows that margarine and other 'partially hydrogenated vegetable oil' products are actually worse for us than butter or other saturated fats..."* - Dr. Jonathan V. Wright, *Nutrition and Healing*

An informative book on chelation is *The Chelation Answer* by Dr. William Campbell Douglas, publisher of the *Second Opinion* medical newsletter. The book is available for $19.45 ($16.95 plus $2.50 shipping and handling) from Second Opinion Publishing Co., 1350 Center Dr., Suite 100, Dunwoody, GA 30338. The telephone number is (800) 728-2288.

For congestive heart failure or other heart problems, Co-enzyme Q10 can work wonders. The body produces it, but weak hearts are often quite deficient. Magnesium is another essential for good heart health and at least the amount in the formula at the beginning of the section should be taken daily along with calcium. Vitamin E intake should be increased to 1200 IU with about half natural and half synthetic. Taurine and Acetyl-L-Carnitine are also very helpful.

Ginkgo biloba has a number of preventative properties for cardiovascular disease.

For individuals with elevated plasma homocysteine levels, which is a good indicator of the possibility of a sudden heart attack, they should take 1000-2000 mcg of folate a day to reduce these levels according to the *New England Journal of Medicine*. In addition, B6 and B12 also help metabolize homocysteine. L-arginine in large doses can stop the process of atherosclerosis or prevent its start. This is the kind of information that the FDA tries to suppress by raiding health food stores and doctor's offices. Dr. Jonathan Wright has been proclaiming that a deficiency in folic acid is a major cause of heart attacks for many years and has been saving his patients' lives with it, despite bitter persecution by the FDA. Now conventional medicine is starting to catch up to him, as more and more studies confirm what he has been saying. I wonder how may tens of thousands of people would be alive today if this information had been widely available to the public instead of being suppressed.

INFERTILITY – MALE

Male sperm counts have dropped fifty percent. Rates of detection of testicular and prostate cancer have doubled. This is thought to be due to increased exposure to pesticides and other toxic chemicals that didn't exist until after World War II, according to E. Douglas Whitehead, M.D. (Urologist).

Fertility can be increased by the supplementation mentioned in the beginning of this section. Other factors in reducing sperm counts temporarily are hot baths or tight underwear. Sperm are very sensitive to temperature.

KIDNEY STONES

Two physicians at Harvard Medical School treated 149 patients who had one or more kidney stones each year. They gave them 300 mg of magnesium oxide and 10 mg of vitamin B6 per day. Before the test they'd had 194 kidney stones the previous year. After they started taking the magnesium, they only had 16 stones. Those results make magnesium one of the best performing "drugs" in any area of medicine.

LONGEVITY

Everyone wants to live a long time, or most people do. Unfortunately, with the virtual explosion of diseases, that may not be so pleasant. Much of the body's deterioration has to do with the reduction in production of various chemicals and hormones and the deterioration of brain cells. The body's reduced production of DHEA, dopamine, melatonin, serotonin, neopenephrene, testosterone, and other chemicals cause many of the problems associated with aging. DHEA is kind of a master chemical from which other chemicals and hormones are made. It can drop to only twenty percent of normal by age eighty. DHEA supplementation is safe and effective. A DHEA sulfate test can determine how much you are producing and a knowledgeable doctor who specializes in this type of supplementation can tell you how much to take to return your levels to that of a twenty-five-year old, which is the optimum level for adults. L-Deprenyl (Eldepril in the U.S.) will help restore normal levels of dopamine and will supplement to some extent the levels of serotonin and neopeniphrene. It helps regenerate dopamine producing cells that haven't been too badly damaged. L-Deprynly is used in Parkinson's disease, which results in a loss of dopamine. It is widely used in Europe as a life extending drug. Experiments on rats shows that it can prolong normal life expectancy by more than thirty percent.

KH3 and GH3 are also life extension drugs. The effective ingredient is procaine. KH3 is the more stable and preferred form. Procaine helps protect brain cells from the damages of aging among other benefits. Hydergine is another life extension drug widely used in Europe that primarily affects brain chemicals.

Antioxidants also help considerably in prolonging life by preventing cell damage from free radicals. If you are taking the supplements recommended at the beginning of this section, you probably are getting enough of them.

A healthy diet, lots of exercise, and the above drugs could easily extend your life to one hundred. Nothing beats good genes, but these will augment whatever genes you have. If you don't get killed by some drunk driver, drug addict or the FBI, you could live a long and healthy life. If you are interested in this area, you may want to become a member of the Life Extension Foundation listed in the Appendix. They can provide you with stacks of clinical studies and information as well as sources of these drugs.

MS/ALS

Multiple sclerosis and alterial lateral sclerosis are very mysterious diseases whose exact causes are unknown. Typical treatments do little to halt the progression of the disease. Many of the people who develop ALS have had polio as a child. The connection is not known, but there is some correlation.

For some patients, there is one little-known treatment that is proving to be quite effective in many cases. First, all the mercury fillings are removed by a trained mercury-free dentist. They take great precautions to keep the mercury vapor and particles generated by the removal process from getting into the patient's system. Normal removal techniques may be worse than leaving them in. Then, the patient undergoes about twenty chelation infusions to remove the mercury from their system. The connection is not clear, but for many people, complete or substantial remission of symptoms is achieved, and they don't come back.

Fluoride and strong alternating electric/magnetic fields have also been implicated in MS. Smoking, sleeping pills, tranquilizers, zinc and aluminum can aggravate the symptoms. Eliminate milk and yogurt. Don't drink soft drinks, as they contain phosphoric acid or coffee or alcohol. Avoid hydrogenated oils like the plague. Eliminating sugar will help considerably with the fatigue that plagues MS patients.

The following nutritional agents can help relieve the symptoms. EPA, Flaxseed oil, Octacosanol, Pantethine, Vit. C, Squalene, Vit. E, L-Cysteine, L-Glutathione.

Calcium EAP, particularly in the injectable form, has reversed the progress of MS in over sixty percent of those receiving it. Unfortunately, the FDA does everything they can to block the import of it. The tablet form can be obtained in the U.S. Check with the Atkins Center in New York City for more information on MS treatments. Another source of information is the AK Brewer Science Library in Richland Center, Wisconsin.

PROSTATE DISEASE

Prostatisis or an enlarged prostate eventually strikes almost every male if he lives long enough. Prostate cancer is a leading cause of death. Both problems can be significantly reduced by taking saw palmetto, pygeum and zinc. If you are taking the supplement levels at the beginning of this section, you will not need any more zinc as you can get too much. Both saw palmetto and pygeum are widely used in Europe to prevent and treat prostate problems. Saw palmetto shrinks enlarged prostates better than the leading drug with none of the side-effects, which will probably lead the FDA to ban it so the drug companies can sell their more expensive drugs.

ROOT CANALS

If you develop some nagging or debilitating symptoms after having a root canal, you just might be suffering from some very powerful toxins generated by bacteria within the tooth. Very few doctors or dentists are aware of the problem, and may treat you for all kinds of things which accomplishes nothing. Antibiotics do no good. The bacteria can't be reached by the antibiotics as they are deep within the tooth with no blood supply to carry the drugs. The toxins slowly migrate out of the tooth to do their damage. The only recourse is to drill out the tooth and try to eliminate the bacteria, or to pull the tooth. The latter is one of the things that led to the discovery of the problem. People who had puzzling and untreatable health problems suddenly recovered when a tooth with these sealed-in bacteria was pulled for some other reason. Up to one-third of today's cases of degenerative diseases are caused by infected teeth. Included are sinus flare-ups, chronic fatigue, heart disease, arthritis, painful headaches, kidney disease, rheumatism and many more. For all the gruesome details, get a book called *Root Canal Coverup* by Dr. George Meining, which is in the book list in the Appendix.

SORENESS FROM EXERCISE

When you're about to take an extra-hard run or hike, or try out a fall-down sport such as in-line skating, take 400 milligrams of ibuprofen four hours before you start. You'll be far less sore the next day, says a researcher at Texas Women's University, than if you took it right after the exercise. Regular exercise, along with proper warmups and stretching before exercise will help a great deal.

ULCERS

One of the most effective medicines for ulcers is licorice, specifically a form called DGL. It is widely used in Europe because it is more effective than the standard drugs and has none of the nasty side-effects. Now that we know that many ulcers are caused by a bacterium, it sppears that the powerful anti-bacterial properties of licorice may be responsible. Another good treatment for ulcers is *fresh* raw cabbage juice right out of the juicer.

VACCINES

More and more evidence is coming in on the dangers of routine inoculations. Some of these reactions, even deaths, soon after the inoculation are obviously. Now we are finding out that there are delayed results that often don't manifest themselves until much later in life. Some vaccines are contaminated. Others compromise the immune system for unknown reasons. The vast majority of vaccines are for non-serious, almost inconsequential childhood diseases that you are probably better off catching than having inoculations. It may be more uncomfortable, but you will acquire superior immunity without compromising your immune system or having some of the other problems that are showing up. Typically, the medical establishment plays down these problems and states that the results are worth the minuscule risk. Maybe, as long as you or your children aren't the ones that develops the problems.

After a lot of double-talk and denials, it has been admitted by the Defense Department that much of the sickness known as the Gulf War Syndrome was probably caused by the experimental vaccines the troops were given. Some of them got sick before they even got to the Gulf. The same symptoms are showing up in English servicemen who got wholesale vaccinations.

WATER

Many chronic conditions and diseases, including asthma and allergies, can be eliminated or significantly helped by the simple step of drinking eight large glasses of water every day. It is amazing what this can accomplish. This is pure water, not soft drinks, fruit juice, or coffee. The problem is, where do you get pure water? Typical tap water is dangerous to your health. The EPA regularly releases data on the dangerous levels of contaminants in various public water supplies. They don't include chlorine which is deliberately added to the water. Repeated studies have shown how chlorine causes heart disease and other problems in water supplies. Yet, they keep adding it to your body because they can't find anything better to kill bacteria in the water.

Some of the ailments that have been helped or eliminated by just drinking eight glasses of pure water a day for several months include asthma, allergies, chronic fatigue, back pain, and many others.

Fluoride is something else you should not be taking into your body, which is added to many of the water supplies. It is a very powerful poison and is used in rat poisons. When water supplies were first fluoridated, people on dialysis died because of the fluoride. The story of how the aluminum industry found a way of getting rid of their toxic waste, sodium fluoride, by getting states to require mandatory fluoridation is a case study in governmental malfeasance. No level of government has any business forcing you to take any medicine or chemical "for your own good". Fluoride is extremely toxic. People who need large quantities of water, such as construction workers in the summer, can have serious problems from fluoridated water. For the rest of us, the effects may be more subtle. More and more, fluoride is being implicated in some forms of cancer. If it were an accidental additive, it would be banned by the EPA. Many things that are less toxic are banned. If you want the questionable benefits of fluoride for your teeth, topical application can provide it. Don't let it get inside your system.

Where do you get pure water? This is not all that easy. Some ethical bottled water companies can provide it. Only use a reputable company that continuously has their water tested and will provide you with a written copy of the results. Many brands of spring or mineral water aren't much better than tap water, except they usually don't contain chlorine. Make sure the water you drink has been decontaminated with ozone, not chlorine.

There are a number of ways you can get pure water directly. Distilling can deliver pure water if processed properly. However, some experts believe that drinking distilled water can be harmful. Reverse osmosis filters with carbon filters can remove most contaminants from tap water. If they are used with well water, you may need to use pre-filters for rust and sediments and follow with ozone for pathogens. A new ceramic filter with imbedded silver to kill bacteria is a good choice as a final filter.

Try drinking eight large glasses of water each day for two weeks. You will end up going to the bathroom a lot more often. While that may be a bid of a bother, you will be flushing a lot of junk out of your body and hydrolyzing your cells. If some mysterious ailments go away, don't be surprised. This procedure has even cured chronic back pain caused by insufficient water to properly lubricate the spinal disks.

Chapter Fifteen

A Few Final Thoughts

As I walked by into the kitchen this morning, I noticed that our dog Misty's food and water dishes were still in their usual place in the back hall. I need to pick them up and put them away. They aren't needed anymore. A few days ago I had to put our fourteen-year-old German Shepherd to sleep – permanently.

It was not easy. She had been tightly interwoven into the fabric of our lives for a long time. I hope that there are dogs in heaven, but I doubt it. If dogs *were* there, some people would have packs of them following them around. If dogs are there, what about all the other billions of animals? Somehow, I can't picture dinosaurs running around up there, but that's not my problem or decision. God has promised that the whole scene would blow our minds and that is good enough for me.

The dishes started me thinking about the purpose of life, Misty's life in particular, and life in general. To her, we had provided a place to sleep, food, affection, and those yearly trips to the vet, for her own good, that she hated. In return, we received completely unconditional love, acceptance, and loyalty. Her dancing feet and wagging tail whenever we came into sight clearly communicated to us that we were the most important things in her life. I can't help but wonder if God wouldn't appreciate a response like that from us. We certainly appreciate it, even from a dog. Misty also demonstrated on several occasions that she would attack anything that she thought was a threat to one of us, without any regard for her own life.

If God did not exist, what would be the point of life (quantum leap there, but stay with me)? If we are just the result of hopelessly impossible numbers of useful random collisions and combinations, nature (whatever that is) will have played the all-time cruelest joke on all of us. All our work, all our struggles, all our seeking, and all our pain and suffering would be utterly pointless. The hundred million people the Communists have slaughtered trying to form the perfect society, even if they could possibly achieve it, would be only a tragic waste. Eventually, the human race would destroy itself fighting over dwindling resources, long before the sun burns out and everything on earth dies. As big a fan of *Star Trek* as I am, we are not going to find another perfect planet with a younger sun that we can migrate to.

Even the "eat, drink, and be merry, for tomorrow we die" approach doesn't work. There is never enough "merriment" to hold off the nagging sense of despair and meaninglessness of it all for long. Besides, if there is no life after death, then after a few short years you will cease to exist, and can't remember the good times anyway. Also, most of the merriment most people choose to indulge in is destructive, and leads to even fewer years of life.

Fortunately, we have not been left in the pits of ultimate despair. God has chosen to reveal Himself to us and His purposes for it all in the Bible. Some people, like Josh McDowell, who is a lawyer, set out to prove the Bible false. What he found was overwhelming, irrefutable evidence that not only does God exist, but Jesus is exactly who the Bible says He is. If that sort of pursuit interests you, you can find all the details in McDowell's books, starting with *Evidence That Demands a Verdict*.

Some people have started out trying to disprove that Jesus was indeed the promised Messiah, but just a prophet or a good teacher, only to run into over three hundred specific prophecies about His first coming. The probability for anyone to fulfill even ten of these prophecies would reach beyond impossible, and with each additional one that impossibility grows not by a factor of one but by another huge factor of multiplication. Try keeping that up for three hundred more such multiplications. The numbers become incomprehensible, long before you get anywhere near three hundred.

Many people have tried to discredit the Bible by saying it was a bunch of fairy tales and that people, groups, and even kingdoms, such at the Hittites, that are mentioned in the Bible never existed. According to the scoffers, the Hittites never existed. There was no trace or mention of such a people in all of history. The Bible is obviously false. Well, there wasn't any evidence until some archeologist took the information from the Bible and started digging where they felt the information led them. Guess what? The Hittite kingdom was every bit as large and powerful as the Bible said it was.

All of that is very helpful, I suppose, but the reason I came to believe in God is that the Bible made sense. It was internally consistent over thousands of years and many writers. It gave reason and purpose to a marvelous universe that *obviously* had to be designed by an incredible intelligence. Psalm 19:1 says that the heavens declare the glory of God and the firmament shows His handiwork. So does the genetic code, perhaps even more so. In apparent reference to that psalm, Percival Lowell, one of the world's most famous astronomers said: "Surely, the undevout astronomer must be mad."

No one in their right mind would pick up a fine Swiss watch on a beach and concluded that it was the result of countless years of sand being washed around by the tides. The universe, and our genetic code, are hundreds of orders of magnitude more complex than any watch. Genetic code is "written" in a way that could not possibly have happened by accident. One hundred billion monkeys pounding on typewriters for one hundred billion years could not produce one page of Shakespeare, much less his complete works. The randomness is too great. Linear order requires intelligent design. Laws, even the laws of physics, require a lawgiver to design and enforce the laws.

We can never know an infinite God through our own efforts, much less become one ourselves. He is the Creator; we are the creatures. Yet, He has made a way to adopt us into His family as His children. We can only know Him as He chooses to reveal Himself to us. He has done that in the Bible. That is how I came to know about Him and to believe in Him, but He wants much more than that. He wants close loving fellowship with each of us. That's why we were created and given this carefully crafted planet to live on. Being the source of perfect love is rather useless, if there is no one to love and no one to return that love by his/her own *free* choice. That's why He provided an answer for all our screwups. To believe in Him is not enough for Him; He wants us to know Him, love Him, and enjoy His company. I am not talking about religion here. I am talking about fellowship. Most religions drive or draw people away from God and leave them struggling for some way to reach

Him or some understanding of Him. Why bother, when all we have to do is accept what He has freely given and enjoy it?

I don't just believe in God, I know Him personally. If that sounds incredibly egotistical or ridiculous, then you don't really understand "our *Father* who is in Heaven." He reveals much about Himself in the Bible, and we can know a lot about Him, His nature, His character, and His capabilities from it. But we still don't know Him personally as Father. That only comes by accepting Him into our lives, accepting Jesus as the needed sacrifice to provide for reconciliation with a perfect God, and walking through life with Him in continuing fellowship.

Every day, I get the incredible privilege to know Him better. As the old hymn says, "He walks with me and He talks with me..." That talk is not a one-way conversation. He speaks in numerous ways. If you are a Christian and have only a one-way conversation, He wants much more for you. It is hard to get to know anyone very well without having conversations with them.

I have innumerable proofs of God's existence and His constant care in my own life. They would only be somebody's stories to you, but they are irrefutable, real experiences to me, including immediate miraculous healing of intractable physical problems. I mention that only because there are hundreds of millions of people who can give the same testimony, including some of the most intelligent and gifted people who ever lived. They can't all be deluded.

I wonder how God feels when He not only gives us life in the first place, but a planet specifically designed, to extremely tight tolerances, to provide us with everything we need if we are willing to work a little bit. In return, we ignore Him. Our dog did a lot better than that. Then, He demonstrates *His* unconditional love for us by sending *His* son to die in our place as a sacrifice, to take the punishment we deserve for our indifference, disobedience, and rebellion so we can still have fellowship with Him who is perfect in every respect. We still ignore Him. However, being perfect love, He still keeps trying, calling us to turn to Him and respond to His love towards us.

He has a purpose for my life, and He has a purpose for your life. Unless we seek to find and fulfill that purpose, we will never be satisfied. We can make billions of dollars, like Howard Hughes, and still end up a pathetic prisoner of our own fears, as he did. On the other hand, Mother Teresa has few of this world's goods, by her own choice. Yet, she is completely at peace with herself and with God. She enjoys close daily fellowship with Him. So do I. So can you. You don't have to be a Mother Teresa, or much of anything else, but only willing to come to Him. If you seek after Him, you will find Him. He is not far from any of us.

"If a man gains the whole world, but loses his own soul, what benefit is it to him." (Mark 8:36). It is very unfortunate that Hitler, along with millions of others, never understood that. For me, there is little on earth comparable to just spending time in fellowship with Him. Weird? Don't knock it until you've experienced it. Out of that fellowship, I have no fear of death. I don't particularly look forward to the process of dying, but I *know*, not just believe, that I will be far better off then. I would like to stick around a while longer to watch over my kids as they get married, to see what happens to them and to help if I can. But, I also look forward to that day when Jesus will escort me into my Father's presence, not because of *anything* I did or did not do, but because of what He did. Anyone can have the same assurance. All you have to do is to come to Him with enough humility and common sense to know that He is God and you are not.

What has all of that to do with a dead dog? Probably not much. But it does lead me to the following. To Misty: wherever you are, if you are, thank you for giving us so much joy. Thank you for being such a splendid example of unconditional love and acceptance. You have helped me to better understand this kind of love which God has for us and to know and understand Him a little more. There are five people who loved you and who also will live forever by the grace and mercy of God. Therefore, at least memories of you will live forever in their hearts. Your life was not in vain!

How about your life? Are you of no more consequence than a dead twig as the evolutionists claim? Are all your work, struggles, and achievements just so much dust in an accidental universe? Is there really no meaning to anything? I know better. So can you. In all of life, there is one paramount question: Where will you spend eternity? Do you *know*? You can.

Appendix

References for additional information

Listed below are a number of sources for your further inquiry. While all the material is valuable, there are five books in particular that I recommend that you read to start.

1. *The Power of Total Perspective*, R. E. McMaster, Jr., A.N. International, Inc. P.O. Box 84901, Phoenix, AZ 85071

2. *Toward A New World Order*, Don McAlvany, Western Pacific Publishing Co. P.O. Box 84900, Phoenix, AZ 85071

3. *Dumbing Us Down-The Hidden Cirriculum of Compulsory, Schooling,* John Taylor Gatto, New Society Publishers, 4527 Springfield, Ave. Philadelphia, PA 19143

4. *The Complete Works of Pat Robinson, - The New Millennium-The New World Order-The Secret Kingdom,* Pat Robertson, Inspirational Press, 386 Park Avenue South, New York, N.Y. 10016

5. *Alternative Medicine-The Definitive Guide*, Burton Goldberg Group Staff, Future Medical Publishers, 10124 18th St. C.T.E., Puyallup, WA 98371, (800) 275-2606

BOOKS

A Philosophical Enquiry, Edmund Burke, Oxford University Press, 200 Madison Ave. Oxford, NY 10016

A Survivor's Guide to Home Schooling, Luanne Shackelford and Suzan White, Crossway Books, 1300 Cresent St. Westchester, IL 60187

A Symposium on Creation, Donald W. Patten, Baker Book House, P. O. Box 6286, Grand Rapids, MI 49516

American Political Writing During the Founding Era, Charles S. Hyeman and Donald S. Lutz, Liberty Press, 7440 N. Shadeland, Indianapolis, IN 46250

Architects of Conspiracy, William P. Hoar, Western Islands, 395 Concord Ave. Belmont, MA 02178

Armageddon, Appointment with Destiny, Grant R. Jeffery, Doubleday, 1540 Broadway, New York, NY 10036

Bypassing Bypass, Elmer Cranton, Hampton Roads Publishing, 976 Norgolk Square, Norfolk, VA 23502

Conspiracy: A Biblical View, Gary North, Dominion Press, P.O. Box 8204, Ft. Worth, TX 76124

Darwin's Enigma, Luther D. Sunderland, Master Books, P.O. Box 1606, El Cajon, CA 92022

Declare War on Our Lousy Government, Wayne Green, WGI, 70 Route 202 North, Peterbourough, NH 03458

Democracy in America, Alexis De Tocqueville, Richard D. Heffner, New American Library, 245 Fifth Ave. New York, NY 10016

Dr. Wright's Guide to Healing Nutrition, Dr. Jonathon Wright, Keats Publishing, P.O. Box 876, New Caanan CT 06840.

Economics in One Lesson, Henry Hazlitt, Crown Publishing Group, 201 E. 50th St. New York, NY 10022

Environmental Overkill. Dixy Lee Ray. Harper Collins Publishers Inc. 10 E. 53rd Street, New York, NY 10022

Euro-Quake. Daniel Burstein. Simon and Schuster, Rockefeller Center, 1230 Avenue of the Americas, New York, NY 10020

Financial Guidance, Dr. James McKeever, Omega Publications, P.O. Box 4130, Medford, OR 97501

Food Allergies Made Simple, Phyllis Austin, Newlife Books, 9297 Siempre Viva Road, San Diego, CA 92173

Free Speech Or Propaganda: How the Media Distorts the Truth, Marlin Maddoux, Huntington House, P.O. Box 53788, Lafayette, LA 70505

From Freedom to Slavery, Gerry Spence, St. Martin, 175 Fifth Ave. New York, NY 10010

Government by Emergency, Gary North, American Bureau of Economic Research, P.O. Box 8204, Ft. Worth, TX 76124

How to Help Your Child Survive and Thrive in Public School, Cliff Schimmels, Flemming H. Revell, P.O. Box 6287, Grand Rapids, MI 49516

In The Beginning, Walter T. Brown, Creation-Life Publishers, San Diego, CA 92115

It Is Their Right, James M Bulman, Gateway Publications, P.O. Box 6295, Greensboro, NC 27405

Miseducation: Preschoolers at Risk. David Elkind. Alfred A. Knopf, 201 E. 50th St. New York, NY 10022

NEA: Propaganda Front of the Radical Left, Sally D. Reed, National Council for Better Education, 1800 Diagonal Road, Suite 635, Alexandria, VA 22314

NEA: Trojan Horse in American Education, Samuel L. Blumenfeld, Paradigm Company, c/o Research Publications, P.O. Box 39850, Phoenix, AZ 85069

New Lies for Old, Anatoly Golitsyn, Soundview Publishing, 1350 Center Drive, Dunwood, GA 30338

New Money Or None, Willard Cantelon, Logos International, P.O. Box 11929, Jacksonville, FL 32239

None Dare Call It Conspiracy. Gary Allen and Larry Abraham. Double A Publications. Suite 403, 18000 Pacific Highway South, Seattle, WA 98188

None Dare Call it Conspiracy-Part II, John Stormer, Liberty Bell Press, P.O. Box 32, Flourissant, MO 63032

On Liberty and Other Essays, John Stuart Mill, Oxford University Press, 200 Madison Ave. Oxford, NY 10016

Outdoor Survival Skills, Larry Dean Olsen, Pocket Books, 1230 Avenue of the Americas, New York, NY 10020

Oxygen Threrapies: A New Way of Approaching Disease, Ed McCabe, Energy Publications, Morrisville, NY 13408

Proofs of a Conspiracy, by John Robinson, Western Islands, P.O. Box 8040, Appleton, MI 54913

Red Cocaine: The Drugging of America, Joseph Douglass, Soundview Publications, 1350 Center Drive, Dunwoody, GA 30338

Red Horizons: Chronicles of a Communist Spy Chief, Ion Pacepa, Regnery Publishing, 422 First Street, S.E. Washington, DC 20003

Rights of Man, Thomas Paine, Penguin Books, 375 Hudson Street New York, NY 10014

Root Canal Coverup Exposed, George E. Meinig, Bion Publishing, P.O. Box 10, Ojai, CA 93024

Ruling Class: Inside the Imperial Congress, Eric Felten, Heritage Foundation, 214 Massachusetts Avenue N.E. Washington, DC 20002

Rush To Armageddon. Texe Marrs, Tyndale House Publishers, P.O. Box 80, Wheaton, IL 60189

Scientific Creationism, Henry M. Morris, Creation-Life Publishers, San Diego, CA 92115

Setting Limits-Constitutional Control of Government, Lewis K. Uhler, Regnery Gateway, 1130 17th Street, N.W. Washington, D.C. 20036

The Anti-Federalist Papers, Ralph Ketchum, Penguin Books, 375 Hudeon Street, New York, N.Y. 10014

The Battle for the Family, Tim LaHaye, Flemming H. Revell, P.O. Box 6287, Grand Rapids, MI 49516

The Battle for the Public Schools, Tim LaHaye, Flemming H. Revell, P.O. Box 6287, Grand Rapids, MI 49516

The Birth of the Republic, Edmund S. Morgan, University of Chicago Press, 5801 Ellis Ave. Chicago, IL 60637

The Captive American, Lee Brandenburg, Hampton Books, 333 W. Santa Clara Street, #1212, San Jose, CA 95113

The Chelation Answer by Dr. William Campbell Douglas, Second Opinion Publishing Co., 1350 Center Dr., Suite 100, Dunwoody, GA 30338

The Clinton Chronicles Book, Patrick Matrisciana, Jerimiah Books, P. O. Box 1800, Hemet, CA 92546

The Collapse of Evolution, Scott M. Huse, Baker Book House, P.O. Box 6287, Grand Rapids, MI 49516

The Coming Economic Earthquake, Larry Burkett, Moody Press, 820 N. LaSalle Blvd. Chicago, IL 60610

The Culture of Disbelief, Stephen L. Carter, Doubleday, 1540 Broadway, New York, NY 10036

The End of Liberalism. Theodore J. Lowi. W. W. Norton & Co., 500 5th Avenue, New York, NY 10110

The Fearful Master, A Second Look at the United Nations, by G. Edward Griffin, Western Islands, P.O. Box 8040, Appleton, WI 54913

The Federalist Papers. Clinton Rossiter. NAL Penguin 1633 Broadway, New York, NY 10019

The Finger Print of God, Hugh Ross, Promise Publishing Company, 2324 N. Batavia Springs, Orange, CA 92665

The Genesis Flood, Henry M. Morris, Baker Book House, P.O. Box 6287, Grand Rapids, MI 49516

The Home Invaders, Donald E. Wildmon, Victor Books, 1825 College Ave. Wheaton, IL 60187

The Ideological Origins of the American Revolution. Bernard Bailyn. The Belnap Press of Harvard University, Cambridge, MA

The Impeached President, Nick Guarino, 1129 E. Cliff Road, Burnsville, MN 55337

The Naked Capitalist, W, Cleon Skousen, Bucanneer Books, P.O. Box 168, Cutchogue, NY 11935

The New Money Survival Handbook, Dr. Ron Paul, 1120 NASA Blvd,, Suite 104, Houston, TX 77058

The Ruling Class-Inside the Imperial Congress, Eric Felten, The Heritage Foundation, 214 Massachusetts Ave. NE, Washington, DC 20002

The Samurai, The Mountie, and The Cowboy, David B. Kopel, Prometheus Books, 59 John Glenn Drive, Buffalo, NY 14228

The Troubled Waters of Evolution, Henry M. Morris, Creation-Life Publishers, San Diego, CA

The Unseen Hand, A. Ralph Epperson, Publius Press, 3100 S. Philamena Place, Tucson, AZ 85730

The Vermont Papers, Frank Bryan and John McClaughry, Chelsea Green Publishing, P.O. Box 428, White River Junction, VT 05001

Thomas Jefferson On Democracy. Saul K. Padover. New American Library, 245 Fifth Ave. New York, NY 10016

Timely and Profitable Help, Hans J. Schneider, World Wide Publishing, P.O. Box 105, Ashland, OR 97520

Toxic Metal Syndrome, Dr. H. Richardson Casdorph and Dr. Morton Walker, Avery Publishing Group, Garden City Park, NY

Tragedy and Hope, Dr. Carroll Quigley, GSG And Associates, P.O. Box 6448, Eastview Station, Rancho Palos, CA 90734

Trashing the Planet, Dixy Lee Ray, Harper Collins Publishers, 10 E. 53rd Street, New York, NY 10022

Trilateralism. Holly Sklar. South End Press, Box 68, Astor Station, Boston, MA 02123

Trilaterals Over Washington, Volumes I and II, by Antony Sutton and Patrick M, Wood, August Corp. P.O. Box 582, Scottsdale, AZ 85252

Two Treatises of Government, John Locke, Mentor Books, 245 Fifth Ave. New York, NY 10016

Understanding the Dollar Crisis, Percy L. Greaves, Jr., Free Market Books, P.O. Box 186, Irvington, NY 10533

Wall Street and the Bolshevik Revolution, Anthony Sutton, Bucanneer Books, P.O. Box 168, Cutchogue, NY 11935

What Are They Teaching Our Children? Mel and Norma Gabler. SP Publications, Wheaton, IL 60187

When the World Will Be As One: The Coming New World Order In the New Age, Tal Brooke, Harvest House, 1075 Arrowsmith, Eugene, OR 97402

COIN DEALERS

R W. Bradford, P.O. Box 1167, Port Townsend, WA 98368 (800) 854-6991

International Collectors Associates, P.O. Box 5150, Durango, CO 81301 (800) 525-9556

Ron Paul & Associates, Inc. 18333 Egret Bay BLVD., Suite 265, Houston, Texas 77058, Precious Metals (800) 982-7070

INTERNET

Newsgroups
alt.conspiracy
alt.med.allergy
alt.med.fibromyalgia
alt.politics.bush
alt.politics.clinton
alt.politics.correct
alt.politics.greens
alt.politics.libertarian
alt.politics.org.batf
alt.politics.org.fbi

alt.politics.usa.constitution
alt.privacy
alt.society.civil-liberties
alt.society.civil-liberty
alt.society.conservatism
info.nra
info.firearms
info.firearms.politics
misc.activism.militia
misc.health.alternative
misc.health.arthritis
misc.health.diabetes
misc.survivalism
rec.guns
talk.politics.guns
talk.politics.libertarian
talk.politics.medicine

FTP SITES

NRA: gopher://GOPHER.NRA.Org
ftp://FTP.NRA.Org

Patriot Archives, ftp://tezcat.com/patriot

WEB PAGES

NRA, http://www.nra.Org
Journal for Patriotic Justice in America,
http://weber.u.washington.edu
Guide to Internet Firearms Information Resources,
http://www.portal.com/~chan/firearms.faq.html

NEWSLETTERS & MAGAZINES

If you write to the newsletters listed below and request of a free copy, many of them will send you one. Others will send you detailed information on the newsletter and often a special deal for new subscribers.

Alternative Medicine Digest, 10124 18th Street, Ct. E, Puyallup, WA 98371

Alternatives, Dr. David G. Williams, MD, Mountain Home Publishing, 2700 Cummings Lane, Karrville, TX 78028

Conservative Chronicle, Box 11297, Des Moines, IA 50340-1297 (800) 888-3039

Criminal Politics, Box 1028, 8045 Zurich, Switzerland

Destiny (Black Conservative Opinion) P.O.Box 19284, Lansing, MI 48901, (800) 545-584

Don Bell Reports, P.O. Box 2223, Palm Beach, FL 33484

Financial Privacy Report, Daniel Rosenthal, Box 1277, Burnsville, MN 55337

Global Gold Stock Report, P.O. Box 4357, Casper, WY 82604

Harry Browne's Special Reports, P.O. Box 5586, Austin, TX 78763 (800) 531-5142

Health & Healing, Dr. Julian Whitaker, MD, Phillips Publishing, 7811 Montrose Road, Potomac, MD 20854

Health Alert, Dr. Bruce West. MD, P.O. Box 22620, Carmel, CA 93922

Health Revelations, Dr. Robert Atkins, MD, Wellness Communications, 1101 King Street, Suite 411, Alexandria, VA, 22314

Howard Phillips' Inside Washington Report, 450 Maple Ave,, East, Suite 309, Vienna, VA 22180

Human Events, 7811 Montrose Rd. Potomac, Maryland 20854 (800) 787-7557

Intelligence Digest, Intelligence International Ltd. 17 Rodney Road, Cheltenham, Glos, GL50 1HX, United Kingdom

Issues & Views (Black Conservative Opinion, economics) P.O. Box 467 New York, N.Y. 10025, (212) 886-1803

Last Chance Health Report, University of Natural Healing, 355 W. Rio Road, Suite 201, Charlottesville, VA 22906

Low Profile: Your Monthly Guide to Privacy and Asset Protection, P.O. Box 84910, Phoenix, AZ 85071

McAlvany Intelligence Advisor, Don McAlvany, P.O. Box 84904, Phoenix, AZ 85071, (800) 528-0559

Money Strategy Letter, P.O. Box 1788, Medford, OR 97501

Myers' Finance Review, P.O. Box 467939, Altanta, GA 31146

National Minority Politics (Black Conservative Opinion) 5757 Westheimer Suite 3-296, Houston, TX 77057, (800) 340-5454

Nutrition & Healing, Dr. Jonathan Wright, MD P.O. Box 84909, Phoenix, AZ 85071

Precious Metals Digest, Soundview Publications, Inc. Suite 100, 1350 Center Drive, Dunworthy, GA 30338

Remnant Review, Gary North, P.O.Box 84906, Phoenix,AZ 85071, (800) 528-0559

Ron Paul Investment Letter, 1120 NASA Blvd,, Suite 104, Houston, TX 77058

Silver & Gold Report, Precious Metals Report, Inc. P.O. Box 109655, Palm Beach Gardens, FL 33410

Second Opinion, Dr. William Campbell Douglas, MD, Suite 100, 1350 Center Drive, Dunwoody, GA 30338

Soldier of Fortune Magazine, P.O. Box 348, Mt. Morris, IL 61054-9817, (800) 877-5207

Straight Talk, P.O. Box 60, Pigeon Forge, TN 37868

The American Spectator, P.O. Box 655, Mt. Morris, IL 61054-8084 (800) 524-3469

The Global Guide, Gary Scott, International Service Center, 3106 Tamiami Trail North, Suite 9F264, Naple, FL 33940

The Holt Advisory, P.O. Box 2923, West Palm Beach, FL 33402

The New American (anti world government, pro freedom) P.O. Box 8040, Appleton, WI 54913, (414)749-3783

The Reaper, R. E. McMasters, Jr. P.O. Box 84901, Phoenix, AZ 85071 (800) 528-0559

The Ruff Times, Howard Ruff, 757 S. Main Street, Springville, UT 84663

The Wall Street Underground, Nick Guarino, 1129 E. Cliff Road, Burnsville, MN 55337

World News Digest, P.O. Box 467939, Atlanta, GA 30346

ORGANIZATIONS

Accuracy in Media, 1275 K st. N.W. Washington, D.C. 20005 (202) 371-6710
Publishes: various books, pamphlets, posters.

Accuracy in Academia, Inc. 4455 Connecticut Avenue, NW, Suite 330, Washington D.C. 20008, (202) 364-4401 Publishes: Campus Report (anti-PC, College PC)

America's Future, 514 Main Street, New Rochelle, N.Y. 10801 (914) 236-6000

American Center for Law and Justice, P.O. Box 64429, Virginia Beach, VA 23467

American Conservative Union, 38 Ivy Street, S.E. Washington, D.C. 20003, (202) 546-6555

American Council for Capital Formation, 1850 K street. N.W.,Suite 400, Washington, D.C. 20006, (202) 293-5811 Fax: (202) 785-8165

American Council on Science and Health, 1995 Broadway, 16th Floor, New York, N.Y. 10023, (212) 362-7044 Fax: (202) 362-4919

American Defense Institute, 214 Massachusetts Avenue, N.E. Washington, D.C. 20002, (202) 544-4704

American Defense Preparedness Association, 1050 17th Street, N.W.,Suite 1200, Washington, D.C. 20036, (202) 331-1389

American Foundation for Resistance International, 1110 Vermont Avenue, N.W. Washington, D.C. 20005, (202) 429-0107 Fax: (202) 293-3414

American Freedom Coalition, 1001 Pennsylvania Avenue, N.W.,Suite 850, Washington, D.C. 20004, (202) 393-1333 Fax: (202) 393-1337

American Freedom Movement, P.O. Box 309, Hwin, PA 15642

American Legislative Exchange Council, 214 Massachusetts Avenue, N.E., Suite 430, Washington, D.C. 20002, (202) 547 4646

American Security Council, 916 Pennsylvania Avenue, S.E. Washington, D.C. 20003, (202) 484-1676 Fax: (202) 543-8255

American Studies Center, 499 South Capital Street, S.W., Suite 417, Washington, D.C. 20003, (202) 488-7122
Fax: (202) 484-1613

Americans for Nuclear Energy, 2525 Wilson Boulevard, Arlington, Va. 22201, (703) 528-4430

Americans for Tax Reform, 1156 15th Street, N.W. Washington, D.C. 20005, (202) 331-0551

Americans for Effective Law Enforcement, Inc. 5519 North Cumberland Avenue, Suite 1008, Chicago, IL. 60656, (312) 763-280   Fax: (312) 763-3225

Americans United for Life, 343 South Dearborn Street, Suite 1804, Chicago, IL 60604, (312) 786-9494   Fax: (312) 341-2656

Atlantic Legal Foundation, 205 E. 42nd Street, New York, N.Y. 10017, (212) 573-1960   Fax: (212) 573-7550

Capital Legal Foundation, 700 E Street, S.E. Washington, D.C. 20003, (202) 546-5533

Capital Research Center, 1612 K Street, N.W., Suite 704
Washington, D.C. 20006, (202) 822-8666   Fax: (202) 785-5634

Cato Institute (Libertarian) 224 2nd Street, S.E. Washington, D.C. 20003, (202) 546-0200 Fax: (202) 546-0728

Cause Foundation (Waco Defense Fund) P.O. Box 1235, Black Mountain, N.C. 28711

Center for Responsible Politics, 1320 19th Street, N.W. Washington, D.C. 20036, (202) 857-0044

Center for Strategic and International Studies, 1800 K Street, N.W., Suite 400, Washington, D.C. 20006, (202) 887-0200

324 - Restoring America

Center for Inter-American Security, 122 C Street, N.W., Suite 710 Washington, D.C. 20001, (202) 393-6622   Fax: (202) 737-7235

Center for Security Policy, 1250 24th Street, N.W., Suite 600 Washington, D.C. 20037, (202) 466-0515   Fax: (202) 466-0518

Center for Constructive Alternatives, Hillsdale College, Hillsdale, MI 49242, (517) 437-7341

Center for the Study of Market Alternatives, 1920 E. Hazel, Caldwell, ID 83605

Center for the Study of Popular Culture, 12400 Ventura, Blvd., Suite 304, Studio City, Calif. 91604, Publishes: Heterodoxy (anti-P.C.)

Center for the Study of American Business, Washington University Campus, Box 1208, 1 Brookings Drive, St. Louis, Mo 63130, (314) 889-5630   Fax: (314) 889-5688

Christian Broadcasting Network, CBN Center, Virginia Beach, VA 23463, (804) 424-7777   Fax: (804) 523-7812

Christian Coalition (lobbying, home school, traditional values) P.O. Box 1990, Chesapeake, VA 23327-1990

Christian Financial Concepts, P.O. Box 100, Gainesville, GA 30503

Christian Legal Defense and Education Foundation, P.O. Box 1088, Fairfax, VA 22030, (703) 818-7150

Christian Rescue Effort for the Emancipation of Dissidents (CREED) 787 Princeton Kingston Road, Princeton, N.J. 08540 (609) 497-0224

Christian Voice, 214 Massachusetts Avenue, N.E., Suite 300 Washington, D.C. 20002, (202) 544-5202   Fax: (202) 544-5213

Citizens Committee to Keep and Bear Arms, 12500 N.E. Tenth Place Bellevue, Wash. 98005, (206) 454-4911
Fax: (206) 451-3959

Citizens Committee against Government Waste, 1301 Connecticut Avenue, N.W., Suite 400, Washington, D.C. 20036, (202) 467-5300

Citizens for Educational Freedom, 927 South Walter Reed Drive, Arlington, Va, 22204, (703) 486-8311

Citizens for America, 214 Massachusetts Ave., N.E., Suite 480, Washington, D.C. 20002, (202) 544-7888   Fax: (202) 543-5502

Citizens for a Sound Economy, 70 L'Enfant Plaza, S.W., East Building, Suite 7112, Washington, D.C. 20024, (202) 488-8200

Clearinghouse on Educational Choice, 927 S. Walter Reed Drive, Suite 927, Arlington, Va 22204, (703) 486-8311

Coalition for Freedom, P.O. Box 19458, Raleigh, NC 27619, (919) 781-4190

Committee for the Survival of a Free Congress, 717 Second Street, N.E. Washington, D.C. 20002, (202) 546-3000
Fax: (202) 546-7689

Committee on the Present Danger, 905 16th Street, N.W., Suite 207, Washington, D.C. 20006, (202) 628-2409

Competitive Enterprise Institute, 233 Pennsylvania Avenue, S.E., Suite 200, Washington, D.C. 20003, (202) 547-1010
Fax: (202) 546-7757

Concerned Women for American, 370 L'Enfant Promenande SW, Suite 800, Washington, DC 20024, (202) 488-7000
Fax: (202) 408-0806

Conservative Caucus, INC. 450 E. Maple Avenue, Vienna, Va 22180 (703) 938-9626

Conservative Leadership Conference, c/o Griffin Communications 713 Park st.,SE, Vienna, VA 22180, (703)255-2211
Fax: (703) 281-6617

Conservative National Committee, 2030 Clarendon Blvd., #305, Arlington, Va 22201, (703) 522-2104

Conservative Network, 444 North Capital Street, N.W. Washington, D.C. 20001, (202) 347-3121   Fax: (202) 737-4405

Conservative Victory Fund, 422 First Street, S.E. Washington, D.C. 20003, (202) 546-833

Council for National Policy, 513 Capitol Court, N.E., Suite 200 Washington, D.C. 20002, (202) 675-4333

Council on Domestic Relations (counter group of the Council on Foreign Relations, anti-new world order) P.O. Box 3362, Springfield, IL 62708

Eagle Forum, Box 618, Alton, IL 62002, (618) 462-5415

Education and Research Institute, 800 Maryland Avenue, Washington, D.C. 20002, (202) 546-1710

Empower America, 1776 I Street NW, Suite 890, Washington, D.C. 20006-3700, (800) 332-2000

Ethics & Public Policy Center, 1030 15th Street, N.W., Suite 300, Washington, D.C. 20005, (202) 393-1200   Fax: (202) 408-0632

F.R.E.E. Box 8616, Waco, TX 76710

Family Research Council, 700 13th Street, N.W.,Suite 300, Washington, D.C. 20005, (202) 393-2100   Fax: (202) 393-2134

Fisher Institute, 11601 Audelia Road, Suite 153, Dallas, TX 75243 (214) 343-4736

Focus on the Family, Colorado Springs, CO 80995

Foreign Policy Research Institute, 3615 Chestnut Street, Philadelphia, Pa 19104, (215) 382-2054   Fax: (215) 382-0131

Foundation for Economic Education, 30 South Broadway, Irvington-on-Hudson, N.Y. 10533, (914) 591-7230
Fax: (914) 591-8910

Free Congress Research and Education Foundation, 721 Second Street, N.E. Washington, D.C. 20002, (202) 546-3004

Freedom Alliance, 45472 Holiday Drive, Sterling, VA 22170 (703) 709-6622

Freedom House, 48 E. 21st Street, New York, N.Y. (212) 473-9691

Freeman Institute, National Center for Constitutional Studies, 5288 S. 320 West, Suite B158, Salt Lake City, Utah 84107, (801) 261-1776

Fund for a Conservative Majority, 313 Massachusetts Avenue, N.W. Washington, D.C. 20002, (202) 546-3993

Fund for Objective News Reporting, 422 First Street, S.E. Washington, D.C. 20003, (202) 546-0856

Future of Freedom Foundation, 11350 Random Hills Road, Suite 800, Fairfax, VA 22030, (703) 934-6101   Fax: (703) 803-1480

Gun Owners of America, 1025 Front Street, Suite 300, Sacramento, CA 95814

Gun Owners of America Foundation, 8001 Forbes Place, Suite 102 Springfield, VA 22151, (703)321-8585   Fax: (703)321-8408

Hale Foundation, 422 First Street, S.E. Washington, D.C. 20003, (202) 546-2293

Heritage Foundation, 214 Massachusetts Ave. N.E. Washington D.C. 20002, (202) 546-4400   Fax: (202) 546-8328

High Frontier, 2800 Shirlington Road, Arlington, VA 22206 (703) 671-4411 Fax: 703-931-6432

Hillsdale College and Shavano Institute, Hillsdale, MI 49242 (517) 437-7341 Fax: (517) 437-3923

Hoover Institute, Stanford University, Stanford, CA 94305 (415) 723-1754

Hudson Institute, 5395 Emerson Way, Indianapolis, IN 46226 (317) 545-1000   Fax: (317) 545-9639

Individual Rights Foundation, 9911 W. Pico Blvd. Suite 1290, Los Angeles, CA 90035

Institute for Contemporary Studies, 243 Kearny Street, San Francisco, CA (415) 981-3353

Institute on Religion and Democracy, 729 15th Street, N.W. Suite 900, Washington, D.C. 20005, (202) 393-3200 Fax: (202) 638-4948

Intercollegiate Studies Institute (ISI), 14 S. Bryn Mawr Ave. Bryn Mawr, PA 19010-3275, (800) 526-7022

International Freedom Foundation, 200 9th Street, S.E., Suite 300, Washington, D.C. 20002, (202) 546-788   Fax: (202) 546-5488

International Society for Individual Liberty, 1800 Market Street, San Francisco, CA 94102

Landmark Legal Foundation, 1006 Grand, 15th Floor, Kansas City, Mo 64106, (816) 474-6600   Fax: (816) 474-6609

Legal Affairs Council, Freedom Plaza, 14018-A Sully field Circle, Chantilly, VA 22021, (703) 378- 2231

Libertarian Party of Texas, 1716 W. Anderson Lane, Austin, TX 78757, (800) 422-1776

Life Amendment Political Action Committee, Stafford, VA 22554

Life Extension Foundation, P. O. Box 229120, Hollywood, FL, 33022

Lincoln Institute for Research and Education, 1001 Connecticut Avenue, N.W. Washington, D.C. 20036, (202) 387-1011

Ludwig Von Mises Institute, 851 Burlway Road, Burlingame, CA 94010,

Madison Center for Educational Affairs, 1112 16th Street, N.W. Washington, D.C. 20036, (202) 833-1801 Fax: (202) 467-0006

Manhattan Institute for Policy Research, 42 East 71st Street New York,N.Y. 10021, (212) 988-7300 Fax: (212) 517-6758

Media Institute, 3017 M Street,N.W. Washington, D.C. 20007, (202) 298-7512 Fax: (202) 337-7092

Media Research Center, 111 S.Columbia Street, Alexandria, VA 22314, (703) 683-9733

Mountain States Legal Fund, 1660 Lincoln Street, Suite 2300, Denver, CO 80264, (303) 861-0244 Fax: (303) 831-7379

National Alliance of Senior Citizens, 2525 Wilson Blvd. Arlington, VA 22201, (703) 528-4380 Fax: (703) 528-2763

National Association of Scholars, 29 Nassau Street, Suite 250 East, Princeton, N.J. 08542, (609) 683-7878 Fax: (609) 683-0316

National Association of Neighborhood Schools, 1800 W. 8th Street. Wilmington, Delaware 19805

National Association of Evangelicals, 1023 15th Street, N.W., Suite 500, Washington,D.C. 20005, (202) 789-1011 Fax: (202) 842-0392

National Center for Public Policy Research, 300 Eye Street, N.W., Suite 3, Washington,D.C. 20002, (202) 541-1287

National Center for Policy Analysis, First Interstate Plaza, 12655 N. Central Expressway, Suite 720, Dallas, TX 75243, (214) 386-NCPA Fax: (214) 386-0924

National Center for Constitutional Studies, 5288 South 320 West, Suite B-158, Salt Lake City, Utah 84104, (801) 261-1776

National Citizens Action Network (NCAN), P.O. Box 10459, Costa Mesa, CA 92627, Publishes: Family Protection Scoreboard (714)850-0349 Fax:(714)662-3952

National Conservative Political Action Committee, 618 South Alfred Street, Alexandria, VA 22314 (703) 548-0900 Fax: (703) 836-2413

National Conservative Foundation, 618 South Alfred Street, Alexandria, Va 22314, (703) 548-0900 Fax: (703) 836-2413

National Council on Family Relations, 3989 Central Avenue, N.E., Suite 550, Minneapolis, MN 55421, (612) 781-9331

National Council for Better Education, 108A South Columbus Street, Alexandria, VA 22314, (703) 684-4404

National Defense Foundation, 228 South Washington Street,Suite 230, Alexandria, Va 22314, (703) 836-3443 Fax: (703) 836-5402

National Forum Foundation, 107 Second Street, N.E. Washington, D.C. 20002, (202) 543-3515   Fax: (202) 547-4101

National Humanities Institute, 214 Massachusetts Avenue, N.E., Suite 470, Washington, D.C 20002, (202) 544-3158

Appendix - 331

National Journalism Center, 800 Maryland Ave. N.E. Washington, D.C. 20002, (202) 546-1710

National Legal Center for the Public Interest, 1000 16th Street,N.W., Suite 301, Washington, D.C. 20036, (202) 296-1683 Fax: (202) 293-2118

National Pro-Life Political Action Committee, 2525 Wilson Boulevard, Suite 200, Arlington, Va 22201

National Rifle Association, P.O.Box 96916, 1600 Rhode Island Ave. N.W. Washington, D.C. 20090-6916 (202) 828-6000 Fax: (202) 861-0306

National Right to Life Committee, 419 7th Street, N.W., Suite 402, Washington, D.C. 20004, (202) 626-8800 Fax: (202) 737-9189

National Strategy Information Center, 1730 Rhode Island Ave.,Suite 601, Washington, D.C. 20036, (202) 429- 0129 Fax: (202) 659-5429

National Tax Limitation Committee, 201 Massachusetts Ave. N.E., Suite C7, Washington, D.C. 20002, (202) 547-4196 Fax: (202) 543-5924

National Taxpayers Union, 753 Maryland Ave. N.E. Washington, D.C. 20002, (202) 543-1300

National Wilderness Institute, 1001 Prince Street, Suite 100, Alexandria, Va 22314, (703) 836-7404

Pacific Legal Foundation, 2700 Gateway Oaks Drive, Suite 200, Sacramento, CA 95833, (916) 641-8888

Pacific Research Institute for Public Policy, 177 Post Street, Suite 500, San Francisco, CA 94108, (415) 989-0833 Fax: (415) 989-2411

Project Vote Smart / Center for National Independence in Politics 129 NW 4th Street, #204, Corvallis, OR 97330, (503) 754-2746 Fax: (503) 754-2747

332 - Restoring America

Public Advocate, 6001 Leesburg Pike, Suite 3, Falls Church, VA 22041, (202) 546-3224

Public Affairs Council, 1019 19th Street, N.W. Washington, D.C. 20036, (202) 835-8343

Public Service Research Council, 1761 Business Center Drive, Suite 230, Reston, VA 22090, (703) 438-3966   Fax: (703) 438-3935

Reason Foundation, 2716 Ocean Park Boulevard, Suite 1062, Santa Monica, CA 90405, (213) 392-0443

Republican Liberty Caucus (Libertarian) 1717 Apalachee Parkway, Suite 434, Tallahassee, FL 32301, (904)878-4464

Revolution Times (anti-new world order, CFR)P.O. Box 6756, Huntington Beach, CA 92615, (714) 921-7284

Ron Paul & Associates, Inc. 18333 Egret Bay BLVD.,Suite 265, Houston, Texas 77058, (713) 333-4888 / (800) 766-7070 {800-Ron-Paul}

Second Amendment Foundation, James Madison Bldg. 12500 N.E. Tenth Place, Bellevue, WA 98005-2538

Selous Foundation, 325 Pennsylvania Ave., N.E. Washington, D.C. 20003, (202) 547-6963

The American Cause (educational foundation,conferences) 6862 Elm St. Suite 210, McLean, VA 22101, (800) 65Cause   Fax: (703) 827-0592

The American Enterprize Institute, 1150 17th Street, N.W. Washington, D.C. 20036, (202) 862-5800

The Ayn Rand Institute, 4640 Admiralty Way, #715 Marina del Rey, CA 90292,

The Conservative Caucus, 450 Maple Ave, E, Suite 309, Venna, VA 22180

The Foundation for Economic Education, Irvington-on-Hudson, NY 10533.

The Fund For American Studies, 1526 Eighteenth Street N.W. Washington, D.C. 20036, (800-741-6964)   Fax: (202) 986-0390

The Gordon Liddy Letter, 800-(call G Man)

The John Birch Society, P.O. Box 8040, Appleton, WI 54913, (414) 749-378

The Leadership Institute, 8001 Braddock Road, Suite 502, Springfield, VA 22151, (703)321-8580 (800) 827-LEAD   Fax: (703) 321-7194

The Rockford Institute, 934 North Main Street, Rockford, IL 61103 (815) 964-5053   Fax: (815) 965-1826

The Rutherford Institute, International Headquarters, P.O.Box 7482, Charlottesville, VA 22906, (804) 978-3888 For Legal Assistance, "Defending Religious Liberty and the Family"

The Wildlife Legislative Fund of America (WLFA) 801 Kingsmill Parkway, Columbus, OH 43229-1137 (614)888-4868 (alternative to radical environmentalism)

Today's Family, Route 2, Box 656, Grottes, VA 24441

U.S. Alliance for Freedom Fighters, 1420 E. Missouri, Suite 105, Phoenix, AZ 85014, (602) 248-8174   Fax: (602) 248-8243

U.S. Justice Foundation, 2091 E. Valley Parkway, Escondido, CA 92025, (619) 741-8086

U.S. Taxpayers Party, 450 Maple Avenue East. Vienna, VA 22180

United Conservatives of America, 7777 Leesburg Pike, Suite 500, Falls Church, VA 22043, (730) 356-0440

334 - Restoring America

Washington Legal Foundation, 1705 N Street, N.W. Washington, D.C. 20036, (202) 857-0240   Fax: (202) 857-0049

Young America's Foundation, 110 Elden St. Herdon, VA 22070, (703) 318-9608

Young Americans for Freedom (YAF) 14018-A Sullyfield Circle, Chantilly, Va. 22021, (703) 378-1178

Quantity	Description	Price	Total
	Restoring America	$15.00	
	AIDS: A Broader Approach to Treatment	$10.00	
1	Computer-based Home Schooling Information	Free	- - - - -
	Restoring America Newsletter*	$29.00 per yr	
	Postage and Handling	Free	- - - - -
	Total		

*Regular price is $49.00 per year. Purchasers of this book receive a $20 discount on a new subscription.

Make checks payable to Technology Management, Inc.

Send to: Technology Management, Inc.
     1106 Rayburn Court
     Mahomet, IL 61853